33130

ABE

Palm Oil

Critical Reports on Applied Chemistry Volume 15

Palm Oil

edited by F.D. Gunstone

Published for the Society of Chemical Industry by
John Wiley & Sons
Chichester · New York · Brisbane · Toronto · Singapore

Library of Congres Cataloging-in-Publication Data:
Palm oil.
  (Critical reports on applied chemistry; v. 15)
  Includes index.
  1. Palm-oil. I. Gunstone, F.D. II. Society
of Chemical Industry (Great Britain) III. Series.
TP684.P3P35 1987   664'.3   86-23426

ISBN 0 471 91335 9

British Library Cataloguing in Publication Data:
Palm oil. — (Critical reports on applied
  chemistry, ISSN 0263-5917; V.15)
  1. Palm-oil
  I. Gunstone, F.D. II. Society of Chemical
Industry III. Series
  665'.35   TP684.P3

ISBN 0 471 91335 9

Printed and bound in Great Britain

Contents

Editor's introduction ix

1 **Past and prospective world production and exports of palm oil** 1
   S. Mielke

2 **Growth and production of oil palm fruits** 11
   B.J. Wood

3 **Extraction of crude palm oil** 29
   J.H. Maycock

4 **Refining and fractionation of palm oil** 39
   F.V.K. Young

5 **End uses of palm oil** 71
   Human food 71
   S.A. Kheiri

   Animal feed 84
   R.I. Hutagalung

   Industrial uses 92
   R.J. de Vries

Index 99

List of Contributors

R.J. de Vries   *Acidchem (Malaysia) Sdn. Bhd, PO Box 123, Butterworth, Penang, Malaysia*

F.D. Gunstone   *Department of Chemistry, The University, St Andrews, Fife, Scotland LKY16 9ST*

R.I. Hutagalung   *Department of Animal Sciences, Universiti Pertanian Malaysia, Serdang, Selangor, W. Malaysia*

S.A. Kheiri   *Palm Oil Research Institute of Malaysia, Banghi, Selangor, W. Malaysia*

J.H. Maycock   *Palm Oil Research Institute of Malaysia, Banghi, Selangor, W. Malaysia*

S. Mielke   *Oil World Publications, ISTA Mielke and Company, PO Box 90 08 03, 2100 Hamburg 90, West Germany*

B.J. Wood   *Ebor Research, Sime Darby Plantations, PO Box 202, Batu Tiga, Selangor, W. Malaysia*

F.V.K. Young   *Vernon Young Consultant Limited, 67 Freshfield Road, Formby, Liverpool, L37 3HL*

Editor's introduction

The world production of oils and fats is around 60 million tonnes per year. Most of this (about 80 per cent) is used for human food and the remainder is divided between animal feed (about 6 per cent) and oleochemicals (about 14 per cent). This last category is likely to grow because of the increasing interest in using renewable resources in place of non-renewable petrochemicals. Oils and fats are mainly of vegetable origin (72 per cent) with the balance from animal or marine sources. In the recent past soyabean, palm, sunflower and rape have all shown increased production but palm now shows the greatest rate of increase and one estimate suggests that it will be equal to soya (the present leader) in about twenty years time.

This rapid growth in the availability of palm oil, and of its fractionated products palm stearin and palm olein, is reflected in increased R and D activity among producers and users of these products. It is timely therefore to review the subject of palm oil. An account of the past, present and prospective production of palm oil is followed by discussions on the growth and production of oil palm fruits and on the extraction of crude oil from these fruits. Thereafter, there is a review of the refining and fractionation of palm oil and the end uses of palm oil are discussed in three sections: human food, animal feed and industrial uses. This review is confined to the oil extracted from the endosperm (palm oil) and does not cover the distinct and valuable oil from the kernel.

Palm oil production is dominated by activity in Malaysia: this is reflected in the location of the contributors, to whom I offer thanks on behalf of myself and the readers of this review.

F.D. Gunstone

1 Past and prospective world production and exports of palm oil

S. Mielke

1.1   Introduction... 1

1.2   Factors leading to dynamic growth 1

1.3   Future prospects.. 7

1.1 Introduction

Going back more than two decades, we find that palm oil production was static for many years. In the 1950s and 1960s commercial world production stagnated at 1.3–1.4 million tonnes per year. As a result its share of total world production and trade of all major oils and fats had a tendency to decline continuously. In 1968 palm oil production accounted for only 3.8 per cent of the world total (compared with 4.8 per cent in 1958) and exports for 9.2 (13.5) per cent. It was a rather boring commodity, at least for analysts and forecasters.

The dynamic growth of the industry did not begin until 1970, although preparation for it had already started in the 1950s. In the fifteen years up to 1984 world production increased by as much as 325 per cent. During the same period the production of its two keenest competitors, soyabean oil and rapeseed oil, rose by only 157 and 200 per cent respectively.

Table 1.1 reveals that the growth is almost entirely confined to two countries, Malaysia and Indonesia. During the past fifteen years Malaysian production increased 10.5 times and its share of the world total rose to 60 per cent from 24 per cent in 1969. Indonesian production rose 5.5 times and its share to 17 per cent compared with 13 per cent fifteen years ago. The output of all other countries taken as a group has not even doubled, so their share of the world total declined to 23 per cent from 63 per cent in 1969. This happened even though the percentage rate of increase in some smaller producing countries such as Colombia, Papua New Guinea and Solomon Islands was higher than that of Indonesia.

1.2 Factors leading to dynamic growth

It is important to know the factors behind the dynamic growth of the oil palm industry during the past fifteen years because such knowledge will be helpful in getting an idea of the future prospects. The major factors have been the following:
1. Strong government policies of diversification in Malaysia since the 1960s and in

1

Table 1.1 World production of palm oil (1000 t)

Country of production	Jan–Dec 1969	Jan–Dec 1974	Jan–Dec 1975	Jan–Dec 1976	Jan–Dec 1977	Jan–Dec 1978	Jan–Dec 1979	Jan–Dec 1980	Jan–Dec 1981	Jan–Dec 1982	Jan–Dec 1983	Jan–Dec 1984
Benin[a]	13*	12	15*	23	11	10	12	16	17	13	9	9
Cameroon[b]	52	62*	55*	54*	59*	61*	60*	70*	71	75*	72*	80*
Ivory Coast[c]	38	139	146	144	118	151	132	182	155	160	148	174
Nigeria	288*	465*	381*	528*	438*	397*	389*	433*	363*	342*	366	295*
Zaire[d]	200	146	145	129	105	100	95*	98*	92	91	79	76*
Brazil	12	7*	7*	10*	12*	15*	16*	12*	15*	17*	18	21*
Colombia	20	51	51	50	52	67	71	74	80	87	102	118
Ecuador	4*	11	11	22*	25	29*	34*	40*	49*	58*	66	72
China, PR	70*	74*	78*	80*	81	82*	85*	84*	80*	85*	89*	96*
Indonesia[e]	189	351	411	434	497	525	600	691	742	838	900	1084
East Malaysia[f]	26	104	121	131	129	145	155	179	179	261	235	308
West Malaysia	326	942	1137	1261	1484	1640	2033	2397	2645	3253	2783	3408
Solomon Is.	—	—	—	5	7	11	13	14	18	19	20	20
Papua New Guinea	—*	18*	25*	28*	26*	30	36	35*	46*	77*	81	114*
Other countries	228*	263*	274*	248*	258*	275*	284*	294*	305*	323*	355	389*
Total	1466	2645	2858	3145	3302	3538	4015	4619	4858	5700	5321	6264

[a] Output of SONICOG and SOBEPALH.
[b] Output of Socapalm, Parastatals CDC and Palm Oil only.
[c] Output of Sodepalm-Palmivoire-Palmindustrie only; other production, mostly for subsistence, is about 15 000–20 000 t annually.
[d] Industrial output only.
[e] At estates only.
[f] Sabah and Sarawak.
*Denotes estimated figure.

Indonesia since the late 1970s. As Malaysia felt that its reliance on timber, tin and rubber was detrimental for its economy, the government decided to promote research and investment in palm oil. The result can be seen in Table 1.2. The area devoted to palm oil began to expand sharply in the 1960s. The expansion gained momentum in the 1970s and continued into the 1980s. During the 20 years ending 1983 the area planted with oil palm in all Malaysia increased seventeenfold from 77 000 hectares to as much as 1.28 million ha.

In Indonesia the expansion of oil palm area was relatively moderate in the 1960s and 1970s but is more rapid this decade as the government wishes to lessen its dependence on mineral oil for earning foreign exchange. During the current five-year plan oil palm area is targetted to increase to 1.31 million ha by the end of 1988 compared with 400 000 ha at the end of 1983. Almost one-third of the expansion is to take place in Sumatra, the traditional growing area, another one third in Riau and most of the remainder in West Kalimantan, where nucleus smallholder estates, similar to those existing in Malaysia, have begun to be established. Although the extremely ambitious plan is unlikely to be fulfilled, the annual expansion of oil palm area could reach and finally exceed that of Malaysia (in terms of hectareage) from the middle of this decade onwards. A more realistic estimate of total planted area is probably 1.0 million ha by the end of 1988 and 1.1 million ha a year later (Table 1.3).

2. Improvements in oil palm varieties and cultivation and processing technology. The first big breakthrough was the development of the *tenera* variety which boosted yields by some 25 per cent. The next was the development of the fractionation process which opened large new markets in tropical and subtropical countries. The third step was the introduction of the pollinating weevil *Elaeodobius kamerunicus* which has meant a cost saving of M$120–150 per hectare and probably also some increase in yields, as the yield development towards the end of 1984 has indicated. The latest breakthrough has been in tissue culture, the realization of which is just beginning in Malaysia, where 1000–2000 ha were planted at the end of 1984. It is expected to bring a boost in yields by another 25 per cent. This, as the boost by the *tenera* palms, is on top of the gradual increase in yields resulting from a variety of improvements in breeding, cultivation and processing techniques which have been going on continuously.

3. High profitability of oil palm growing. This results from the above two factors plus the ideal climate and the generally good soils available in Malaysia, Indonesia and other countries of South-East Asia. Oil palms produce much higher yields and are much more profitable than other oil crops and other competing crops in the tropics. In West Malaysia, for instance, the crude palm oil yield per hectare increased from 2.8 t in 1966 to 4.1 t in 1982. In addition, about 0.4 t of palm-kernel oil and 0.55 t of meal are produced per hectare. This compares with the normal US soyabean yield of 2.0 t, equivalent to only 0.36 t of oil and 1.6 t of meal. The normal rape seed yield in the EEC is now about 2.5 t, yielding at best 1.0 t of oil and 1.45 t of meal.

Table 1.2 Area, yields and production of palm oil in Malaysia

Year	Areas as of 31 December Total (1000 ha)	Year	Mature (1000 ha)	Average mature area	Yield per hectare (t)	Production (1000 t)
(a) West Malaysia						
1963	71.0	1966	65.9	65.8	2.83	186
1964	83.2	1967	75.1	74.4	2.91	217
1965	97.0	1968	92.5	84.8	3.12	265
1966	122.7	1969	117.1	108.2	3.01	326
1967	160.4	1970	141.2	134.1	3.00	402
1968	198.4	1971	169.5	170.4	3.23	551
1969	239.0	1972	208.1	190.8	3.44	657
1970	270.1	1973	250.3	231.4	3.19	739
1971	293.9	1974	290.3	269.4	3.50	942
1972	348.5	1975	348.4	310.5	3.66	1137
1973	411.9	1976	401.7	362.4	3.48	1261
1974	500.0	1977	490.7	418.9	3.54	1484
1975	568.6	1978	536.8	494.8	3.31	1640
1976	629.6	1979	601.4	557.1	3.65	2033
1977	691.7	1980	700.7	634.5	3.78	2397
1978	755.5	1981	750.7	719*	3.68*	2645
1979	830.5	1982	830*	786*	4.14*	3253
1980	906.6	1983	920*	873*	3.19*	2783
1981	996.5	1984	1020*	963*	3.54*	3408
1982	1048.0	1985F	1070*	1047*	3.57*	3740*
1983	1105*	1986F	1125*	1095*	4.02*	4400*
1984	1200*	1987F	1220*	1170*	4.02*	4700*
1985F	1290*	1988F	1310*	1250*	4.12*	5150*
1986F	1380*	1989F	1400*	1340*	4.14*	5550*
1987F	1470*	1990F	1490*	1430*	4.06*	5800*
1988F	1560*	1991F	1580*	1520*	4.25*	6460*
1989F	1650*	1992F	1670*	1610*	4.35*	7000*
(b) East Malaysia						
1963	6.0*	1966	6*	6*	1.48*	8.9
⋮						
1976	85.0	1979	80*	70*	2.22*	155.5
1977	90.1	1980	85*	81*	2.21*	179.1
1978	97.5	1981	90*	86*	2.08*	179.2
1979	108.3	1982	100*	95*	2.75*	261.3
1980	116.7	1983	110*	105*	2.25*	235.4
1981	122.8	1984	120*	115*	2.68*	308
1982	137.8	1985F	135*	127*	2.48*	315*
1983	150*	1986F	152*	140*	2.71*	380*
1984	160*	1987F	162*	155*	2.81*	435*
1985F	170*	1988F	172*	165*	2.79*	460*
1986F	190*	1989F	193*	182*	2.99*	545*
1987F	210*	1990F	213*	200*	3.00*	600*
1988F	230*	1991F	233*	220*	3.22*	710*
1989F	250*	1992F	253*	240*	3.33*	800*

F = forecast. *Denotes estimated figure.

Table 1.3 Area, yield and production of palm oil in Indonesia

Year	Areas as of 31 December Total (1000 ha)	Year	Mature (1000 ha)	Average mature area	Yield per hectare (t)	Production (1000 t)
1967	—	1970	89.9*	88*	2.46*	216.5
1968	—	1971	93.7*	92*	2.70*	248.4
1969	—	1972	99.9*	96*	2.81*	269.4
1970	133.5	1973	113.8*	106*	2.74*	290.0
1971	139.3	1974	135.2*	124*	2.83*	351.1
1972	152.3	1975	143.8*	139*	2.96*	411.4
1973	158.1	1976	152.4*	148*	2.93*	433.9
1974	182.2	1977	163.3*	157*	3.17*	497.7
1975	190.0	1978	191.1*	176*	2.98*	525.0
1976	203.8	1979	207.6*	198*	3.03*	599.9
1977	233.4	1980	231.4*	219*	3.16*	691.0
1978	245.8	1981	242.8*	236*	3.14*	741.8
1979	265.3	1982	265.0*	250*	3.35*	837.7
1980	294.1	1983	295.0*	278*	3.24*	900.4
1981	328.1*	1984	300*	296*	3.66*	1084.0
1982	380*	1985F	350*	324*	3.42*	1107*
1983	441	1986F	430*	385*	3.30*	1270*
1984	542*	1987F	520*	470*	3.35*	1575*
1985F	650*	1988F	630*	570*	3.25*	1850*
1986F	770*	1989F	750*	690*	3.30*	2275*
1987F	880*	1990F	860*	800*	3.25*	2600*
1988F	980*	1991F	960*	910*	3.36*	3060*
1989F	1080*	1992F	1060*	1010*	3.40*	3430*

F = forecast. *Denotes estimated figure.

A recent West Malaysian study has shown that, already, at palm oil prices of M$700–850 per tonne, f.o.b. mill, and a fresh fruit bunch yield of 18.5 t, the profit per hectare is M$1100–1695 compared with only M$340–860 for rubber at a price of 200–265 cents per kilo.

From August 1983 to July 1985 the domestic price for crude palm oil in Malaysia was above M$1000, with the exception of August 1984 when it averaged M$980. It has thus far exceeded the production costs of M$500, which is the average of a range of M$350–600. This includes depreciation on the mill as well as the original investment in planting, though that is largely taken care of by tax advantages.

4. Ample availability of land suitable for oil palm growing in Malaysia, Indonesia and other countries of South-East Asia. This land consists mostly of virgin forest and partly also of land previously used for the production of rubber and other crops less profitable than oil palm.

5. The firmness of the US dollar. This made both oils and meals more expensive in those consuming countries that are dependent to a large degree on imports of the oil seeds or the products themselves, but has affected the consumption of oilseed meals, especially soya, even more. This is because a number of valuable substitutes (especially grains but also corngluten feed) is

available in place of oilseed meals while virtually no substitutes are available for seed oils used for food. Butter is too expensive to be used effectively in significantly increasing amounts as a substitute while the production of other edible animal fats is increasing only slowly. Thus the main effect of the US dollar firmness has been a significant slowing down of the world consumption of oilseed meals, especially soya. Consequently, the production of seed oils, again especially soya, has slowed down significantly. This is the main reason why the increase in world production of soyabean oil, the growth leader up to the early 1970s, has slowed appreciably since 1981 (see Figure 1.1).

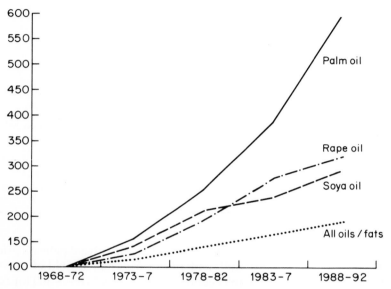

Figure 1.1 Index of world production of palm oil, soya oil, rape oil and all oils/fats (index 1968–72 = 100).

Palm oil thus has quite naturally taken the place of soyabean oil as the growth leader in the production of oils and fats. In fact, had it not been for the big increase in palm oil production in the past fifteen years, the two acute shortages that developed in 1974 and again in 1984 would have been even worse and the price rises correspondingly higher. It is a myth that palm oil production has been rising too sharply and has pushed aside other oils and fats, especially soyabean oil, to gain larger market shares. In reality soyabean oil and all other oils and fats, taken as a group, would no longer have been in a position to cover the increase in demand at reasonable prices.

There is a strong probability that at least four out of the five factors mentioned above will remain in force during the next five to eight years. Since the middle of 1983 some of the factors have become even stronger than they were in the five to eight years before. This refers above all to the enhanced profitability resulting from the much higher prices and the cost saving since the introduction of the weevil. In the years to come it will be further improved by the higher yields to be expected from tissue culture. At the same time government help in research and financing is continuing — in Indonesia it will reach unprecedented levels in the next five years or so.

Only the currency factor has reversed and will cease to be a factor favouring palm oil production.

1.3 Future prospects

The rapid growth of palm oil production can be expected to continue during the next eight years throughout South-East Asia, but especially in Malaysia and Indonesia. Malaysian production is likely to more than double, probably reaching 7.4 million t by 1992. The expected development of planted and mature area as of the end of each year and the average mature area harvested in the course of each year together with the yield per hectare and total production are shown in Table 1.2. The forecasts assume normal weather conditions and the continuation of the normal three-year yield cycle, i.e. two years up and in the third year down. This is due to the desire of the trees to rest after two 'on-years'.

Some Malaysian forecasters are considerably less optimistic, chiefly on the grounds that increasing labour shortage may cause a more pronounced slowing of the expansion than has been assumed in this review. There has been talk about a shortage of labour for a number of years, but the expansion has hardly slowed. Considering the experience of the past 20 years or so, the high profitability can be expected to be maintained at least during the next six to eight years, taken as a whole. This should, as hitherto, help solve the labour shortage through increased harvest mechanization and/or higher wages and fringe benefits. On the other hand, the data in Table 1.2 shows that *some* slowing of the expansion has been assumed to take place chiefly as a result of the labour factor.

Indonesian production is expected to increase even more sharply, and should more than treble to 3.4 million t in 1992 compared with about 1.07 million t in 1984. This will be due exclusively to the much sharper rate of increase in area, even though the area expansion assumed in this analysis is far below the official target. The increase in the palm oil yield per hectare is expected to be considerably less pronounced than in West Malaysia owing to the increasing share of young and thus lower-yielding trees (Table 1.3).

Outside the two major producing countries, too, the increase in production is expected to continue at almost the same rate as in the recent past. Much of the increase is in the smaller countries of South-East Asia. Most of the rest is in South and Central America. After a continued decline or at least stagnation up to 1986, some recovery is expected also for Africa as the decline in Nigeria should then be reversed.

Table 1.4 shows the development of annual world production and exports in five-year averages from 1968 up to 1992. It shows that the rate of production

8   S. Mielke

Table 1.4 World production and exports of palm oil; annual five-year averages (1000 t)

	1968–72	1973–77	1978–82	1983–87	1988–92
Producing country					
Nigeria	387*	432*	385*	330*	366*
Indonesia[a]	222	397	679	1187*	2643*
East Malaysia[b]	37	112	184	335*	623*
West Malaysia	440	1113	2394	3806*	5992*
Other countries	690*	773*	904*	1292*	1770
Total	1776	2827	4546	6950	11394
Exporting country					
Nigeria	11	7	1*	—*	—*
Indonesia[c]	192	348	357	405	1360*
East Malaysia[b]	36	108	178	325*	600*
West Malaysia	426	1016	2052	3400*	5250*
Other countries[d]	379*	505*	798*	1195*	1450*
Total	1044	1984	3386	5325	8660

[a]At estates only.
[b]Sabah and Sarawak.
[c]Crude, refined and stearin.
[d]Includes re-exports (especially large from Singapore, but also from EEC).
Source: Oil World, Hamburg, with slight revisions by the writer.
*Denotes estimated figure.

growth is expected to be 10 per cent annually in the first five years and almost 13 per cent in the last five.

However, because of the irregular yield cycle and the uncertainty of the weather, forecasts of five-year (or three-year) average palm oil production are more realistic than any forecast for a single year so far ahead. However, assuming normal weather in 1990 and 1991, and an on-year yieldwise in both Malaysia and Indonesia, world production can be expected to reach 13.6 million t in 1992 compared with 6.26 million t in 1984. During the same period world gross exports should increase from 4.6 to 9.5 million t.

The increase in world exports is expected to slow to 12.5 per cent annually during the five years ending 1992 from 14.6 per cent in the five years ending 1987. This is chiefly due to the very sharply increasing domestic consumption of Indonesia and, to some extent, also of Malaysia and other countries. The domestic disappearance in both countries, but above all of Malaysia, includes growing amounts used in manufacturing ghee and other secondary products which are exported (but not included in the exports of crude and processed palm oil). Table 1.5 shows the annual development of exports from 1976 to 1984.

Since the second half of the 1970s a sharply increasing share of world exports of palm oil has taken place in some processed form. In 1983 total world *net* exports reached 3.49 million t. Out of that total as much as 2.9 million t, or 83 per cent, was in processed form, 2.85 million t being from Malaysia. Most of the *net* exports of crude oil of 0.59 million t originated from Indonesia, Papua New

Table 1.5 World exports of palm oil (1000 t)

Exporting country	Jan–Dec 1969	Jan–Dec 1976	Jan–Dec 1977	Jan–Dec 1978	Jan–Dec 1979	Jan–Dec 1980	Jan–Dec 1981	Jan–Dec 1982	Jan–Dec 1983	Jan–Dec 1984
EEC	28	97	109	94	89	121	112	93	123	131
Benin[a]	12	11	6	1	3	11	3P	3P	3*	2
Cameroon	6	6	9	9	6	14	4	12	5	5
Ivory Coast	2	92	79	75	50	96	63	58P	53	49
Nigeria	23	3	1*	3	–*	–*	–*	–*	–*	–*
Zaire	125	40	21	10	–	10	6	4	2	7*
Indonesia[b]	179	406	405	412	354	511	206	302	407	247
East Malaysia[c]	25	128	124	144	155	169	162	258	254	290
West Malaysia	331	1207	1304	1370	1746	2108	2345	2693	2803	2858
Singapore	113	178	277	299	492	679	423	502	420	763
Other countries	26	67*	73*	73*	75*	75*	161*	263*	181*	286*
Total	870	2234	2406	2490	2970	3794	3486	4189	4251	4638

[a]Crude oil only.
[b]Crude, refined and stearin.
[c]Sabah and Sarawak.
*Denotes estimated figure. P = preliminary official figure.

Guinea, Solomon Islands and Africa. (This disregards the exports of reconstituted crude palm oil ('cocktail') from Singapore.)

The overwhelming importance of Malaysia in world export trade can be seen from the fact that in 1983 its net exports accounted, at 2.94 million t, for 84 per cent of total world net exports.

2 Growth and production of oil palm fruits

B.J. Wood

2.1	**Introduction**	11
2.2	**The palm and its fruit**	12
2.2.1	Botany and genetics	12
2.2.2	*Elaeis oleifera*	15
2.2.3	Cloning	15
2.3	**Growing the palm**	16
2.3.1	Environments	16
2.3.2	Establishment	17
2.3.3	Productivity and spacing	17
2.3.4	Thinning	19
2.3.5	Upkeep	19
2.3.6	Recycling organic wastes	20
2.3.7	Pests and diseases	21
2.3.8	Pollination	21
2.4	**Exploitation and economics**	22
2.4.1	Harvesting standard and cycle	22
2.4.2	Harvesting technique	24
2.4.3	Planting cycle	24
2.4.4	Yield cycles and forecasting	24
2.4.5	Organization of plantings	25
2.5	**Conclusion**	25
2.6	**Acknowledgements**	26
2.7	**References**	26

2.1 Introduction

The oil palm produces edible oil (palm and kernel oil) in bigger quantities per unit of land than any other plant, and productivity is increasing; however, despite its commercial importance, it is not a well-known plant. This chapter gives a background to its agriculture, emphasizing topics of current activity or probable future development, particularly relating to processing and the product. More extensive information is available in such texts as that by Hartley (general)[1],

11

Turner and Gillbanks (practical planting)[2] and Corley *et al.* (research)[3], as well as in the proceedings of a series of international conferences held in Kuala Lumpur since 1968.

2.2 The palm and its fruit

2.2.1 Botany and genetics

The principal oil palm of commerce is *Elaeis guineensis*, in the same botanical tribe (Cocoinae) as the coconut. It is a typical palm, with a crown of pinnate fronds on a thickened vascular stem, with a single bud in the crown centre (Figure 2.1). It produces about 24 fronds per year (the newest opened is known as No. 1), with an inflorescence primordium in each axil. These become macroscopically visible, on dissection, about 30 months before emergence. Each inflorescence contains the primordia for both male and female flowers (monoecious). The sex that develops is determined about 17–25 months before emergence. This is evidently governed by the conditions that prevailed at around that time. The harsher they are, the more likely is maleness, with genetic variation affecting the response of individual palms, further influenced by their earlier production (endogenous cycle). In very severe conditions, inflorescences may be aborted entirely, which explains why not every axil develops one, and how yield may be influenced so far in advance.

Inflorescences can be seen from outside from about 2 months before anthesis (in frond Nos. 12–16 approximately). The male (see Figure 2.1) comprises a whorl of spikelets on a central stalk, each bearing 600 to 1500 small yellow flowers. Pollen is shed over 2–4 days. The female inflorescence becomes the fresh fruit bunch (FFB) of commerce. It has a similar basic structure to the male but with less flowers. Fertilized flowers become fruit (or fruitlets) which develop and ripen over about 6 months. Some fruit develop without fertilization (parthenocarpic) but generally these are unimportant.

The mature bunch contains from a few hundred to a few thousand fruit (Figure 2.2), depending on genetic and environmental factors and palm age. They range from perhaps 5 kg in young poor palms to as much as 40 kg in 15-year-old palms in a good state. Individual fruit are generally in the range 8–20 g. They comprise exocarp (skin), mesocarp (which contains palm oil and water in a fibrous matrix), endocarp (shell) and kernel (the seed), which contains oil and residual meal. A guide to relative quantities is given in Figure 2.2, but variations are large.

Wild and semiwild groves in Africa contain mainly a thick-shelled variety of palm called dura. This was also used in the early plantations. Those in the Far East largely descended from four palms brought to Indonesia in 1848 (the Deli dura). These had good productivity and, by dura standards, a thin shell. Better tended wild groves have a proportion of thinner-shelled palms called *tenera*, which have

Figure 2.1 A 7-year-old oil palm. Some facing fronds have been cut off to show more clearly (1) the central still-unopened leaves (spears) and (2) a male inflorescence in anthesis. Developing fruit bunches can be seen and the arrangement of fronds in eight spirals is shown by the old frond butts on the stem.

more mesocarp and hence more oil. A few tenera were grown in Sumatra from about 1941, but little was known of the control of shell thickness.

An important discovery in the 1940s was of the genetic basis for this difference in shell thickness. The tenera is a cross between dura palms and a third type, pisifera. This latter has a very thin shell or is shell-less, and often the female inflorescences are sterile. It was long thought to be merely an aberrant type.

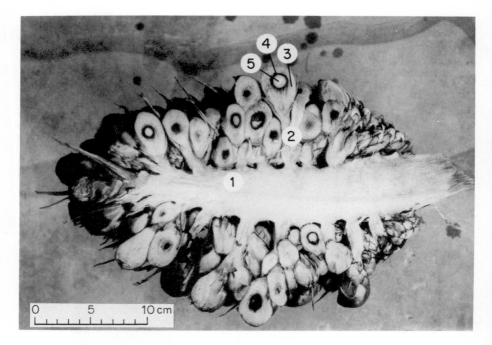

Figure 2.2 A *ca.* 7 kg tenera bunch halved to show (with approximate 'typical' ratios) (1) the stalk plus (2) spikelets (35–40 per cent of bunch) and fruits (60–65 per cent of bunch), comprising (3) the mesocarp (M) (76–78 per cent of fruit; 65–70 per cent of dry to wet M; 78–82 per cent of oil to dry M), (4) the shell (11–14 per cent of fruit) and (5) the kernel (9–12 per cent of fruit).

Virtually all commercial plantings since the 1960s have been with tenera palms, and breeding programmes have developed lines of dura and pisifera as high-grade parents for the cross.

Factory oil extraction ratios (o.e.r.) of 15–19 per cent are obtained with Deli dura (the factory o.e.r. percentage is about 2.5 less than the oil to bunch ratio (O/B) due to unavoidable milling losses). Tenera materials of the late 1960s and early 1970s gave about 20–22 per cent o.e.r. in their early years of bearing, while recent materials may give up to 26 per cent. Wild groves, of course, would not achieve these ratios. There is some suggestion (by Ebor Research in an unpublished work) that o.e.r. of tenera declines with age.

At least nine breeding programmes exist [4]. Those operated by commercial organizations tend to stress production of high-grade seeds from currently available parent materials. Those of organizations of wider answerability concentrate on developing the potential for future generations. Prospection in the original home of the palm is currently taking place[5] to ensure both preservation of genetic material in the face of increasing alienation of wild sites and a wider basis of selection.

2.2.2 *Elaeis oleifera*

This is a close relative of the oil palm from South America (they are sometimes referred to respectively as the African and American oil palms). In recent years it has been of special interest, although there is not as yet any extensive practical use. The oil has similar fatty acids but a higher proportion of the unsaturated (more liquid) members. The two species hybridize readily and the hybrids themselves are partially fertile, so plantings could be envisaged. The oil type is intermediate (Table 2.1) and by backcrossing to either parent species, individuals with a wider range of oils can be produced. Commercial exploitation could increase the range and versatility of palm oils. Unfortunately, the yield of *E. oleifera* is generally very low, as is its O/B, and the hybrids are also not highly productive. At the moment with plants from seeds, the oil from *E. guineensis* could give a bigger quantity of unsaturated oil by fractionation, apart from residues of more solid oils.

Table 2.1 The range among individual palms of oleic and linoleic acid (as a percentage of the total)[6–8]

Fatty acids	*E. guineensis*	Hybrid	*E. oleifera*
18:1 oleic	35–56	37–60	56–70
18:2 linoleic	5–16	8–17	6–23[a]
Iodine value	45–58	60–72	81–84

[a]Omitting a figure of 2.1 given by Meunier and Hardon[6], which is well outside all other ranges quoted.

2.2.3 Cloning

Nearly all current oil palm plantings are from seeds from crossings. As usual in such cases, there is a wide variation and the overall characteristics of the plantings are dependent on the average from a range of genotypes. Thus, around 60–70 per cent of the yield is usually produced by 25–35 per cent of the palms. Cloning of outstanding individuals would be desirable, as is done with many other tree crops. The reason for the qualifying word 'nearly' in the opening sentence is that, of late, there has been an important development in cloning of oil palms by *in vitro* tissue culture. Excised tissue from a parent plant is grown in a sterile medium with plant hormones. In the broadest terms, by successively manipulating the composition of the media, the tissue will undergo cell multiplication (callus), differentiation into plantlets (embryogenesis) and root initiation. Plantlets are then transferred to soil.

Clonal oil palms are being produced in at least three centres and others are developing the technique, using roots and leaves. Initially the tissues were from seedlings (of unknown performance), but some are now being produced from outstanding mature palms (ortets) selected in the field[9], which is obviously preferable. A few tissue cultured palms were field planted from 1977 onwards and those of the first few years are growing normally and uniformly. Unfortunately, some clones planted more recently from long-productive cultures have shown some flowering abnormalities[10]. The cause of this is not known, other than that it is connected with time and media (perhaps as yet unspecified factors in the composition of the latter). In view of the desirability of field tests of performance before clonal selection is complete, there is optimism that the time to resolving the problem will not unduly delay commercial availability of clonal palms.

The potential yield can be exemplified by two cases. One is from a tenera progeny test[11]. The mean oil yield per palm was equivalent to about 6 t/ha, but 5 per cent of palms produced over 10 t. The second (by Ebor Research) is from 800 palms recorded for 30 months, beginning at 8 years old. The mean yield per hectare per year was equivalent to 24.6 tonne at 27.6 per cent O/B, or 6.8 t of oil/ha, but the five top-ranking palms produced the equivalent of 43–48.5 t of fresh fruit bunches (FFB)/ha, and several palms had O/B over 30 per cent, including two of the five highest yielders. The latter give gave oil equivalents of from 11.3 to 14.3 t/(ha year). Allowing for environmental influence and an element of competitive ability in some of the outstanding individuals, the suggestion of 30 per cent oil yield increase with clonal plantings looks very realistic. In selecting ortets, emphasis will probably be on higher O/B. Desirable secondary characteristics can more easily be incorporated and the oil characters of the hybrids and their offspring (Table 2.1), which have good yield in some individual palms, could become commercially exploitable. Even some of the pure *E. oleifera* palms described by Ooi *et al.*[7] may be suitable for cloning.

Some further caution remains necessary. Oil palm cultures remain embryogenic for longer than do those of most plants (possibly indefinitely)[12], but there is a risk of mutation. Widespread susceptibility to disease (in a single clone) could lead to serious outbreak, and yield fluctuations could be exaggerated by coincidence of individual palm cycles. Polyclonal planting could help to solve problems and much remains to be tested.

2.3 Growing the palm

2.3.1 Environments

The oil palm probably evolved in damp but not water-logged alluvial soils, probably on river banks in Africa. In cultivation it does best on similar alluvial or marine clays or well-textured volcanics, with a high water table but good drainage. It thrives best within 10° of the equator. Plantings outside those limits exist but reports of their productivity are difficult to find. In that zone, the evapotranspiration potential is around 0.5 mm/d, and evenly spaced rain at about 1500–2000 mm/year is optimum. In ideal conditions, yields are above 30 t of FFB/ha with today's planting materials. Poor soil texture or chemical characteristics, a relatively high proportion of gravel to concretions, and compaction all pull down the yield potential, but the palm is reasonably adaptable and 15–30 t/ha may be obtained on less favourable soil when climate is good[13]. Even in a strongly seasonal climate, with 3–4 months dry, yields of 10.5 t of FFB/ha can be obtained[14]. Improvement of soil water relations is generally advantageous—e.g. drainage to keep water table below the surface in flat or low-lying areas and conservation measures in the form of terraces and silt pits on sloping lands.

Oil palm planting most commonly follows jungle, a tree crop (particularly rubber), an old stand of palms or field crops. Clean clearing has been favoured in recent years. It is still necessary where land preparation (drains, terraces, etc.) needs to be undertaken, but the current inclination favours retention of organic materials to the maximum extent possible [15], providing that allowance is made to prevent a build-up of disease, particularly after previous stands of palms.

2.3.2 Establishment

Field plantings are generally with palms in the 12–18 month range, an age partly governed by ease of handling. Currently, the immature period is in the range of 27–36 months, and shortening this would obviously be profitable. A current objective is mechanical handling of young palms in bigger bags in order to put them out when more mature. Young palms do not thrive in shade and a lengthy period of underplanting has an adverse effect, but the initial establishment may be done in some circumstances under the old stand just before it is felled. This also ties in with the concept of retaining organic material. In the Far East this practice carries a risk of transmitting *Ganoderma* disease, but this is minimized if the large tissues of the old palms are completely severed from their roots at felling.

2.3.3 Productivity and spacing

The palm converts carbon dioxide, water and nutrient elements to dry matter (DM) by photosynthesis and this is partitioned into vegetative and reproductive (bunch) tissues. The ratio of total DM to bunch (bunch index, BI) varies according to environmental suitability, the genotype of the palms and competition for available resource[16]. The objective in deciding on spacing is to maximize bunch production per unit of land area. The DM and BI are maximum when a palm has no competition, but then palms are too far apart to maximize production per unit of land area. As they become closer, total DM production increases, while the BI of individual palms decreases with increasing competition. Thus there is an optimum spacing for each environment and range of genotypes[17]. Some examples of productivity are given in Table 2.2.

Determining the best density is complicated because competition increases with age, so in the early years bunch productivity is optimum at a greater density than later. Thus, total lifetime yield is highest at a higher density than that which gives

Table 2.2 Growth rates and partition of assimilates of oil palms[1]

Situation (soil type)	Age	Number of palms per ha	Dry matter production/(ha year)		
			Total[a]	Bunches[b]	Bunch index[c]
Nigeria	10–17	148	18.0	5.4	0.30
Nigeria	20–22	148	19.5	4.3	0.22
Malaysia (alluvial)	6.5–17.5	121	28.5	12.5	0.44
Malaysia (alluvial)	27.5	121	17.0	5.8	0.34
Malaysia (sedentary) — progeny trial:					
Mean	10	138	27.6	12.8	0.46
High BI	10	138	30.8	16.9	0.55
Low BI	10	138	23.2	6.0	0.26
Malaysia (sedentary) — density trial:					
Low density	7	112	21.7	11.8	0.54
Medium density	7	145	28.4	14.9	0.53
High density	7	184	32.3	15.1	0.47
High density	7	227	33.6	13.4	0.40

[a]Known as crop growth rate (CGR).
[b]Bunch DM is about 55 per cent of total FFB weight.
[c]Ratio bunch to total DM weight (BI).

optimum yield during full maturity. Furthermore, financial return received earlier is preferable, which further increases the optimum economic density. Taking these aspects into account, Corley et al.[18] recommend densities in Malaysian conditions (climate not severely limiting) of 150 palms/ha on the best soils to 165 on the worst, with 158 as a 'general' density. Goh[19] notes a continuing decline in optimum density with age, and suggests 136 as a 'general' density. Closer densities may be appropriate where climate is limiting[14]. Height increment is affected by density, but there is only a small difference within any realistic range around the optimum density[16].

Because of the marked variability in the BI between individual palms (e.g. see Table 2.2) recent selection of planting materials has stressed this character rather than high individual yield per palm. In seedling plantings, this seems more likely to be achieved with palms of small total DM production[20], but with the possibility of cloning individual palms, a high BI might be found in big individual palms. Lower density would still give the desired result of a high BI[21] and would be preferable for general management.

The palm, unlike many angiosperm tree crops, cannot adapt to fill an available shape; hence, density appears to be more critical. Since the crown is circular, it best occupies space when planted on an equilateral triangle pattern. On contour terraced land it is important to aim at near-equal spacing, with distances measured horizontally and not along the slope. Despite its relative lack of versatility in this respect, the yield to density curve is still fairly flat around the optimum, which evidently is due to variability among genotypes — different individuals would do best at different densities and conditions. With clonal plantings, this buffer will be lost, so that spacing might well be even more critical.

2.3.4 Thinning

Because of the gradual reduction in the optimum density current at a given age as palms get older, deliberate high-density planting, with later thinning, has sometimes been considered. Trials have not usually been successful, but in most examples either the initial or final density was respectively unrealistically high or low[18]. Any planting for deliberate later thinning should have a final stand on an equilateral triangle pattern and would have to be systematically thinned (i.e. equal space all round, which means a minimum of 50 per cent removal from an initial hexagonal pattern). The calculated optimum density of 136 palms/ha (see Section 2.3.3) shows a current optimum in later maturity of about 100 palms/ha[19]. As 200 palms/ha is not an unrealistic current optimum in early maturity, it may be worthwhile to explore the possibility, in various environments, of planting this and thinning later.

Systematic thinning of an existing mature stand apparently well above its current optimum density is not practicable. Nevertheless, removal of some palms to give equal extra light to the remainder (14 per cent, the centre one in every surrounding hexagon, is the minimum removal on an even planting) is sometimes practised[22]. There is no experimental evidence that this is beneficial. Available estimates suggest that when a palm is lost from a stand, yield compensation by surrounding palms ranges from 90 per cent at 143 palms/ha to 60 per cent at 134 palms/ha[23]. More than 100 per cent compensation may be possible at very high density and the possibility is worth investigating in trials, since there are many extensive plantings well above their current optimum density that are not ready for replanting.

2.3.5 Upkeep

Plantations require regular agricultural attention to maximize productivity. Usually the biggest single cost to the grower is for fertilizers — e.g. 12–32 per cent of operating costs in the Far East[24]. Considering the export of up to 16 t of DM/year, the amount of tissue locked up in the palm, and high leaching losses, this is not surprising. Calculating the economic optimum is complicated, because fertilizers have their effect on yield up to two years later, a time for which prices cannot be forecast.

The needs (assessed by economic response in experiments) vary widely according to environment and crop level, but one or more elements is always likely to become limiting. A yield of 25 t of FFB/(ha year) would require the equivalent of 2–3 kg per palm of ammonium sulphate (AS), a nitrogen (N) source, about 0.5 kg of rock phosphate (RP), 1.2 kg of muriate of potash (MOP), and about 1 kg of kieserite (kies). Requirements in productive soils of low chemical status may sometimes be as high as 8 kg of AS, 2 kg of RP, 8 kg of MOP and 1.8 kg of kies[2, 25].

A cover of non-competitive ground vegetation is usually maintained. Herbicides are generally used to clear access paths and a clean circle at palm bases for collection of loose fruit.

Old fronds need to be pruned. The optimum retention is about 32–40[26]. Excessive removal leads to crop loss, but as palms get tall, old fronds restrict access so bunches are unharvested and, as a consequence, cause more crop loss than would occur from some overpruning (from Ebor Research).

2.3.6 Recycling organic wastes

Organic wastes that arise within the plantation and from the mill help to conserve soil and improve its texture. Optimizing use of these materials is a current concern[27, 28].

The pruned fronds (18–24 per palm per year), which have a dry weight of about 10 t/ha[27] with the equivalent of about 125 kg of AS, 30 kg of Christmas Island rock phosphate (CIRP), 120 kg of MOP and 70 kg of kies, are stacked in heaps to rot down. They are being considered for an alternative use[29], but the nutrient content, and value in limiting erosion[30] and in providing mulched areas where feeder roots concentrate must be considered.

An FFB yield of 25 t/(ha year) gives about 5 t of oil and 1.3 t of kernel. Of the balance 3.4 t is fruit fibre, 5.5 t is empty fruit bunches (EFB), 1.7 t is shell and 0.8–0.9 t goes into the mill effluent as solid content (at *ca.* 5 per cent). The balance is attributable to moisture changes. Currently, the fibre is burned in the boilers to operate the mill and EFB is usually slow burned to produce bunch ash, a high potassium fertilizer whose alkaline nature is especially useful on acid soils. Effluent is still discharged to water courses in most territories, but in Malaysia pollution laws have now restricted this. The most common solution is to return it to the land as fertilizer, in a graduated amount, raw or after anaerobic digestion. Application is between palm rows in flat bunded beds, by sprinklers or other means[31]. A 60 t/ha mill processing 200 000 FFB/year gives about 140 000 t of effluent. At the apparent optimum rate of 5–10 cm of rainfall equivalent per year, this could be applied over 120–240 ha. The fertilizer saving more than covers application costs and there is a yield increase variously estimated at up to 20 per cent. With digestion in closed tanks, biogas can be collected. In the 60 t mill this will be about 200 000 m^3/d, enough to generate 1300 kW/h of electricity[32].

Following the example of effluent, where both energy and nutrient potential is being exploited, Wood and Yusof[33] speculate on a similar possibility for all mill 'wastes'. The energy equivalent of the total output of a 60 t mill is estimated to be 200×10^9 kcal ($\equiv$ 20 million kg of diesel), with the nutrient content equivalent to 1350 t of AS, 1400 t of MOP and 285 t of CIRP.

2.3.7 Pests and diseases

These require constant alertness because, although not generally of regular or seasonal occurrence, sporadic incidences may be devastating. Leaf-eating pests may increase and cause complete defoliation with a crop loss of 50 per cent in the following year and full recovery delayed longer. Diseases have been known to wipe out entire plantations.

There are numerous insect pests or potential pests of different species, but in related groups, among the main regions[34–36]. An outstanding feature is that although they can increase to outbreak numbers in the continuously favourable conditions, mostly they do not. This is due to environmental suppression, in which insect natural enemies are a major factor. Common pests include leaf-eating caterpillars (especially in Asia and South America), a leaf-mining beetle (in West Africa) and a stick insect (in Papua New Guinea, PNG), usually present in small numbers but capable of flaring up drastically on occasion. Integrated pest management (IPM) programmes are widely and successfully practised[37–39].

In Malaysia large populations of rats cause losses estimated at around 5 per cent of the oil if no control is practised[40]. Control can be obtained by a system of six-monthly baiting campaigns[41]. Lately, warfarin resistance has been noted[42], but new ('second-generation') anticoagulant poisons are proving effective in such cases.

There are several pathogenic diseases or symptoms of unexplained cause[43]. Generally, infection is sporadic, but there may be massive losses, e.g. in new regions where precautions are neglected. Ganoderma, which occurs in the Far East and Africa, is primarily a saprophyte, but it can infect living palms if there is a large enough innoculum. It tends to spread in incorrectly performed palm-to-palm replants, and it transmits from foci during maturity. In Africa, vascular wilt (*Fusarium oxysporum*) is the most serious disease and has recently been strongly suspected in Brazil[44]. 'Marchitez' has caused the loss of large plantings in South America — its precise cause (or causes) has not been determined beyond doubt[45, 46].

The best approaches to disease control seem to be avoidance (e.g. reducing the chance of exposure to *Ganoderma* at replanting), maintaining good physiological health and breeding for resistance. Fungicides are only occasionally useful, e.g. against *Curvularia* leaf spot in nurseries.

2.3.8 Pollination

The revelation in the late 1970s that oil palms were pollinated primarily by insects rather than wind explained so many earlier anomalies that it should not have been a surprise, although the belief in wind pollination was previously deeply entrenched. Artificial pollination with various types of puffers was practised in nearly all young palms throughout the Far East, for about 4 per cent of the operating cost[24]; it had to be continued throughout the palm's life in Sabah and PNG. It was never needed in West Africa, even in spells of heavy rainfall that removes pollen from the air[47]. The discrepancy finally led to the investigations that showed that numerous insects were involved in Africa, predominantly six species of the weevil *Elaeidobius*. These were known earlier, but had been thought to be incidental. Once their role was proven, it was clear that they must have coevolved with the palms. Adults transfer pollen, while the grubs develop in male flowers after anthesis[48]. A thrips, *Thrips hawaiiensis* (a tiny insect), was then shown to pollinate the palm in Malaysia and Indonesia. It was at a disadvantage in the exposed conditions of young palms and in very wet times or situations, because pupation took place on the ground[49], which explained variations there. Possibly, except for the chance and long unrecognized adaptation of the thrips to oil palms, the palm might have remained an ornamental in the Far East[50]. In South America, pollination was observed to be by a nitidulid beetle, *Mystrops costaricensis*, with an assumed recent import, *Elaeidobius subvittattus*, predominant in several localities on both the American and African oil palm[51]. No significant indigenous insect pollinators were found in Sabah or PNG. *Elaeidobius kamerunicus* was then imported to Malaysia and PNG[52]. From the first release in 1981 it rapidly spread in the underoccupied biological niche in all territories, most recently Sumatra.

After the weevil, regular assisted pollination became history, but at many mills in Malaysia there was a problem in the early days of stripping the bigger and more tightly packed bunches. This prompted renewed speculation about the wisdom of the import. At a seminar on the weevil in 1984[53] the view crystallized that the problem was due to coincidence of a peak crop with the widespread occurrence of weevil pollination. Mill capacity was stretched, so steam sterilization was often inadequate. Further conclusions made at the seminar were that bunches remain bigger with a less variable proportion of fruit, especially in Sabah and PNG; the kernel extraction ratio has gone up (from about 3–5 to 5–7 per cent); and that the o.e.r. may not have changed much. How things will settle remains to be seen, but one forecast is that total oil yields may not increase (due to a compensating reduction in bunch number for the larger size) while yield fluctuations may be increased.

Among other aspects discussed was that of rats preying on weevil grubs; this may affect populations of one or both, but in what way is not yet evident. Also discussed were how weevil pollination may affect many earlier agronomic conclusions from the Far East, e.g. optimum spacing in a trial in PNG had declined to <100 palms/ha in 10-year-old plams pre-weevil, but returned to 125 afterwards; more fertilizer may be needed to realize higher productivity; the optimum harvesting standard may be different; and so on.

2.4 Exploitation and economics

2.4.1 Harvesting standard and cycle

Bunch production is continuous and partially unsynchronized between palms. Oil starts to accumulate in the bunch from relatively early in its development and builds up rapidly from around 130 days after anthesis (ripening). The quantity maximizes and eventually starts to decline, due to the breakdown of oil into free fatty acids (FFA) and loss of actual bunch tissue. Regular harvesting is required, taking only adequately ripe bunches. It is necessary to have an external sign on which ripeness can be recognized (ripeness standard), and this is generally based on fruit detachment, which commences about 140 days after anthesis.

The average ripeness of FFB from a harvesting round will depend upon the ripeness standard (the stated minimum number of detached fruit) and the interval between rounds. The optimum balance is still the subject of some disagreement in the industry, even within technical circles. Nevertheless, it is one of the most important topics in oil palm cultivation because deviations can lead to vast losses of oil in relation to potential. This lost oil is already paid for, in both money and effort.

The subject received considerable coverage at the above-mentioned 'weevil seminar'[53]. Although exact optima were not agreed, it was generally accepted that a frequency of one to two weeks harvesting is best but that it can be stretched to three. Beyoond that, there is an actual loss of FFB weight, which becomes marked beyond 30 days. Also, FFB cut before fruit starts to detach is definitely below optimum oil content. Evidence was presented that oil builds up for a period after fruit detachment commences, not only in proportion but also by a continuing increase in bunch weight. No final conclusion was drawn on the exact optimum procedure for all cases. It was clear that if every bunch were beyond the start of fruit detachment, on ten-day rounds, oil loss due to unripeness would not be great, but reductions due to this cause or to not collecting loose fruit were shown easily to reach four or more percentage points on the o.e.r. A mill scale comparison confirmed some of the causes of loss of the o.e.r. (Table 2.3). A minimum standard of 'fruit detachment started', when rigidly enforced, was as

Table 2.3 Oil content in large batches of FFB (*ca.* 200 t) from tenera palms 5–12 years old harvested under different conditions[54]

| | Harvesting condition[a] | | Percentage of o.e.r.[c] | | |
Ripeness standard	Days since previous harvest	Ripeness standard of previous harvest	Achieved at factory	Estimated milling loss	FFA (%)
4/kg	20–30	4/kg	21.26	2.68	3.42
1/bunch	2	4/kg	21.24	2.44	2.59
Red	2	1/bunch	19.14	2.66	2.01
10/bunch	7	10/bunch	21.41	1.83	2.11
Red	7	Red	19.64	1.85	2.18
Black	7	Black	17.61	2.54	1.83
4/kg[b]	Variable	4/kg	18.01	2.56	3.62

[a]Fruit/kg of average bunch weight for a field, not for individual bunches. Ripeness standard (i.e. minimum for cutting) at present and previous harvest and time interval determines the *mean* ripeness of the batch.
[b]Loose fruit collection not carefully supervised.
[c]o.e.r. = oil extraction ratio achieved; o.e.r. + estimated loss = total oil/bunch.

good as a 'higher' standard and confirmed the extent of losses due to harvesting FFB before fruit detachment or to not collecting loose fruit.

2.4.2 Harvesting technique

On most plantations these days, bunches (up to 2 or 3 m in height) on younger palms are cut with a chisel, and on taller palms, with a curved knife on a pole. Harvesting becomes more difficult as the palms become 12–15 years old. No mechanical cutting method so far tried seems to offer a realistic economic option. The only recent improvement is in the type of pole used, i.e. aluminium instead of bamboo[55]. Decline in yield occurs in tall palms[56], but how much this is due to age, to enforced overpruning (see Section 2.3.5) or to loss of bunches through difficulty in recognizing them is not clear. Probably all have some effect, though carefully controlled trials suggest no actual decline in FFB production potential, up to 18 years at least. Separate loose fruit collection is often practised to extend and ease the task of the relatively skilled harvesters and to ensure complete collection. Mechanical means are being developed to assist in field collection and draught animals are used, including buffaloes in the Far East.

2.4.3 Planting cycle

As a stand ages, there is a gradual loss of palms and of FFB yield (see Section 2.4.2), and recent findings suggest possibly also a loss of the o.e.r. New planting materials tend to be genetically better. However, the duration of the non-productive period and the capital cost of replanting economically favour retaining the old stand. The optimum cycle is currently taken as 25–30 years, but more precise data on the oil yield differences and production costs may show this to be longer than optimum, especially if current work to shorten the non-productive period leads to practicable results.

2.4.4 Yield cycles and forecasting

In seasonal climates, annual cycles are clear cut, governed by the rain/drought seasons. The amount of rain in the wet season and the total effective sunshine appear to govern the size of the peak some two years later[56]. In less seasonal climates, the position is more complicated. In Malaysia, with reasonably well-distributed rain, crops may vary from 5 or 6 to 12 per cent of the annual total. Influences may be fairly clear, e.g. the single annual peak and trough in the more monsoonal areas, but even without this there are cycles, possibly out of phase between places and often different from year to year. Palm cycles seem to gradually become out of phase, but sometimes a strong influence resynchronizes them and a marked local or countrywide peak (or trough) occurs. The factors governing this are not well understood, physiologically or even empirically.

Influences of extreme weather may be discerned in yield up to two years ahead, but the relationship is far from clear.

2.4.5 Organization of plantings

Wild groves are characterized by low yields, although attention to agronomy can raise them. They are associated with small-scale farming, with the domestic use of oil, and with trading, mainly in kernels.

The palm lends itself to, and benefits by, large-scale management systems[57] with hired labour or on a cooperative system. There is a minimum economic size of planting because of the need for investment in a capital-intensive mill. Economies of scale seem to level off at about 10 000–11 000 ha with a 60 t mill, but small mills can be quite profitable. The nature of the FFB makes parallel control of the plantation and mill preferable. FFB weight is not much affected by low ripeness or absence of detached loose fruit, both of which can reduce the o.e.r. greatly (see Section 2.4.3). The difficulty of recognizing these losses after the FFB is cut from the palm reduces the incentive to the grower to harvest correctly if he sells by weight.

It is difficult to give a fair indication of production costs because of the wide variation in labour and material costs (and needs) between localities. Establishment and immature cost, mainly for planting material, immature nutrition and upkeep, but not land, may be in the region of US$1500 per hectare for three years (with perhaps a 20–25 per cent reduction for each year the immature period is shortened).

About one worker is required for 8 hectares. The production cost of FFB, e.g. in Malaysia where labour is relatively expensive, is estimated at about US$25–60/t of FFB depending on the productivity of the land.

2.5 Conclusion

Current yields are quite remarkable, considering the constraints, and it is not surprising that expansion or resuscitation is taking place in many territories. However, it is important to resist the temptation to continue expanding oil palm growing while allowing loss of oil potential through unharvested fruit, soil erosion and other inefficiencies in land already alienated to the crop.

Incorporation of technological improvement from private and public research and advisory services has featured in oil palm development[58], and further change can be expected. A crucial question concerns adaptation of (and to) clones, including changes to breeding programmes, and exploitation of the full potential of secondary characters. Tissue culture should cut across the slowness of oil palm breeding, so that it can more rapidly approach the ideotype concept of Squire[17]. Anuwar[59] conjectures that this might have increased dry matter production with better partition, slow height gain, resistance to diseases, yield stability, drought tolerance and with options for alternative types of oil with different fatty acid spectra. To this might be added efficient use of inputs, easily cut bunches and clearly recognizable optimum ripeness. Identification of ideotypes requires an objective means of defining the relative importance of the characters, especially if they are somewhat reciprocal. The use of growth-regulating chemicals — to control the relationship of oil content and fruit detachment (ripeness), palm size and bunch characters, particularly to improve oil content in hybrid palms — is under active research[25].

A thriving oil palm industry will help towards raising living standards. Paradoxically, this implies an increase in production costs that can be to the detriment of the industry. This emphasizes the need for maximum efficiency in production.

2.6 Acknowledgements

I am grateful to K.G. Berger, H.L. Foster (PORIM), R.H.V. Corley (Unifield) and colleagues in Sime Darby Plantations for helpful comments on the draft manuscript. Sime Darby Plantations gave permission to publish.

2.7 References

1. C.W.S. Hartley, *The oil palm*, 2nd ed., Longman, London (1977), 806pp.
2. P.D. Turner and R.A. Gillbanks, *Oil Palm Cultivation and Management*, Incorporated Society of Planters, Kuala Lumpur (1974), 672pp.
3. R.H.V. Corley, J.J. Hardon and B.J. Wood (eds.), *Oil Palm Research*, Elsevier, Amsterdam (1976), 532 pp.
4. J.J. Hardon, J.P. Gascon, J.M. Noiret, J. Meunier, G.Y. Tan and T.K. Tam, 'Major oil palm breeding programmes', in *Oil Palm Research*, ed. by R.H.V. Corley, J.J. Hardon and B.J. Wood, Ch. 8, Elsevier, Amsterdam (1976), pp. 109–250.
5. N. Rajanaidu, M.J. Lawrence and S.C. Ooi, 'Variation in Nigerian oil palm germplasm and its relevance to oil palm breeding', in *The Oil Palm in Agriculture in the Eighties*, ed. by E. Pushparajah and P.S. Chew, Vol. I, Incorporated Society of Planters, Kuala Lumpur (1982), pp. 3–18.
6. J. Meunier and J.J. Hardon, 'Inter-specific hybrids between *Elaeis guineensis* and *Elaeis oleifera*', in *Oil Palm Research*, ed. by R.H.V. Corley, J.J. Hardon and B.J. Wood, Ch. 9, Elsevier, Amsterdam (1976), pp. 127–38.
7. S.C. Ooi, E.B. Da Silva, A.A. Müller and J.C. Nascimento, 'Oil palm genetic resources — Native *E. oleifera* populations in Brazil offer promising sources', *Pesq. Agropec. Bras.*, **16**(3), 385–95 (1981).
8. N. Rajanaidu, B.K. Tan and V. Rao, 'Analysis of fatty acid composition (FAC) in *Elaeis guineensis*, *Elaeis oleifera*, their hybrids and its implications in breeding', *PORIM Bull.*, **7**, 9–20 (1983).
9. L.H. Jones, 'The oil palm and its clonal propagation by tissue culture', *Biologist*, **30**(4), 181–8 (1983).
10. R.H.V. Corley, C.H. Lee, I.H. Law and C.Y. Wong, 'Abnormal flower development in oil palm clones', *Planter*, **62**, 233–40 (1986).
11. R.H.V. Corley, 'Clonal planting material for the oil palm industry', *Planter*, **58**, 515–28 (1982).
12. J. Blake, personal communication (1984).
13. H.L. Foster 'The determination of oil palm fertilizer requirements in Peninsular Malaysia. Part II: Effect of different environments', *PORIM Bull.*, **4**, 46–56 (1982).
14. R. Ochs and C. Daniel, 'Research on techniques adapted to dry regions', in *Oil Palm Research*, ed. by R.H.V. Corley, J.J. Hardon and B.J. Wood, Ch. 23, Elsevier, Amsterdam (1976), pp. 315–30.

15. C.F. Loh, 'Current replanting methods for oil palm/rubber — A case for re-evaluation and research', *Planter*, **55**, 114–24 (1979).
16. R.H.V. Corley, 'Effects of plant density on growth and yield of oil palm', *Expl. Agric.*, **9**, 169–80 (1973).
17. G.R. Squire, 'Light interception, productivity and yield of oil palm', Report OP(37)84 General (mimeo), PORIM, Kuala Lumpur, (1984), 72pp.
18. R.H.V. Corley, C.K. Hew, T.K. Tam and K.K. Lo, 'Optimal spacing for oil palms', in *Advances in Oil Palm Cultivation*, ed. by R.L. Wastie and D.A. Earp, Incorporated Society of Planters, Kuala Lumpur (1973), pp. 52–71.
19. K.H. Goh, 'Analyses of oil palm spacing experiments', in *The Oil Palm in Agriculture in the Eighties*, ed. by E. Pushparajah and P.S. Chew, Vol. II, Incorporated Society of Planters, Kuala Lumpur (1982), pp. 393–413.
20. C.J. Breure and R.H.V. Corley, 'Selection of oil palms for high density planting', *Euphytica*, **32**, 177–86 (1983).
21. R.H.V. Corley, personal communication (1984).
22. S.K. Ng, 'Advances in oil palm nutrition, agronomy and productivity in Malaysia', PORIM Occasional Paper 12 (1983), 20 pp.
23. R.H.V. Corley, 'Planting density', in *Oil Palm Research*, ed. by R.H.V. Corley, J.J. Hardon and B.J. Wood, Ch. 18, Elsevier, Amsterdam (1976), pp. 273–83.
24. B.J. Wood, 'Technical developments in oil palm production in Malaysia', *Planter*, **57**, 361–78 (1981).
25. PORIM, Annual Rep., Biology Division (mimeo), Palm Oil Research Institute of Malaysia (PORIM), Kuala Lumpur (1984), 271 pp.
26. R.H.V. Corley and C.K. Hew, 'Pruning', in *Oil Palm Research*, ed. by R.H.V. Corley, J.J. Hardon and B.J. Wood, Ch. 22, Elsevier, Amsterdam (1976), pp. 307–13.
27. K.W. Chan, I. Watson and K.C. Lim, 'Use of oil palm waste material for increased production', *Planter*, **57**, 14–37 (1981).
28. T.K. Hong and Abdul Halim Hassan, 'Recycling waste materials in oil palm plantations — mulching in oil palm', *PORIM Bull.*, **1**, 9–14 (1980).
29. R.N. Muthurajah, 'Potential chemical and industrial uses of oil palm mill bulk waste', in *Proceedings of National Workshop on Oil Palm By-Product Utilization*, PORIM, Kuala Lumpur (1983), pp. 140–7.
30. L.M. Maene, K.C. Thong, T.S. Ong and A.M. Mokhtaruddin, 'Surface wash under mature oil palm', in *Water in Malaysian Agriculture*, Malaysian Society of Soil Science, Kuala Lumpur (1979), pp. 203–16.
31. K.H. Lim, B.J. Wood and C.Y. Ho, 'Optimising nutrient recycling of palm oil mill effluent through different systems of land application on oil palm', in *Proceedings of 5th ASEAN Soil Conference*, ed. by S. Panichapong, C. Niamskul, A. Promprasit and M. Newport, Mapping and Printing Division, Ministry of Agriculture and Co-operatives, Bangkok (1984), pp. D4.1–14.
32. K.H. Lim, S.K. Quah, D. Gillies and B.J. Wood, 'Palm oil mill effluent treatment and utilization in Sime Darby Plantations — The current position', in *Proceedings of Workshop on Current Palm Oil Mill Effluent (POME) Treatment Technology towards DOE Standards*, PORIM, Kuala Lumpur, pp. i–iv; 1–114 (1984).
33. B.J. Wood and Yusof Basiron, 'Planting agricultural land in Malaysia — The oil palm option', *Proceedings of International Seminar on Market Development for Palm Oil Products*, PORLA, PORIM, FAO, MOPGC, PORAM, ed. by Y. Basiron and K.G. Berger, PORIM, Kuala Lumpur, pp. i–viii; 1–302 (1984).
34. P. Genty, R.D. de Chenon, J.P. Morin and C.A. Korytkowski, 'Ravageurs du palmier à huile en Amérique Latine', *Oleagineux*, **33**(7), 325–419 (1978).
35. D. Mariau, R.D. de Chenon, J.F. Julia and R. Philippe, 'Les ravageurs du palmier à huile et du cocotier en Afrique Occidentale', *Oleagineux*, **36**(4), 169–228 (1981).
36. Brian J. Wood, *Pests of Oil Palms in Malaysia and Their Control*, Incorporated Society of Planters, Kuala Lumpur (1968), 204 pp.
37. P. Genty 'Entomological research on oil palm in Latin America', in *The Oil Palm in Agriculture in the Eighties*, ed. by E. Pushparajah and Poh Soon Chew, Vol. II, Incorporated Society of Planters, Kuala Lumpur (1982), pp. 485–98.
38. D. Mariau, 'Insect pests in West Africa', in *Oil Palm Research*, ed. by R.H.V. Corley, J.J. Hardon and B.J. Wood, Ch. 26, Elsevier, Amsterdam (1976), pp. 369–84.
39. B.J. Wood, 'Insect pests in South-East Asia', in *Oil Palm Research*, ed. by R.H.V. Corley, J.J. Hardon and B.J. Wood, Ch. 25, Elsevier, Amsterdam (1976), pp. 347–67.

40. B.J. Wood and S.S. Liau, 'A long term study of *R. tiomanicus* populations in an oil palm plantation in Johore, Malaysia. II — Recovery from control, and economic aspects', *J. Appl. Ecol.*, **21**, 465–72 (1984).
41. B.J. Wood and I. Nicol, 'Rat control on oil palm estates', in *Advances in Oil Palm Cultivation*, ed. by R.L. Wastie and D.A. Earp, Incorporated Society of Planters, Kuala Lumpur (1973), pp. 380–95.
42. C.H. Lee, MD. D. Mustafa, K.G. Soh and E. Mohan, 'Warfarin resistance in *Rattus tiomanicus* (Miller)', *MARDI Res. Bull.*, **11**(3), 264–71 (1983).
43. P.D. Turner, *Oil Palm Diseases and Disorders*, Oxford University Press, Oxford (1981), 280 pp.
44. H.L. van de Lande, 'Vascular wilt of oil palm (*Elaeis guineensis* Jacq.) in Brazil', *Oil Palm News*, **27**, 3 (1983).
45. J.L. Renard, 'Diseases in Africa and South America', in *Oil Palm Research*, ed. by R.H.V. Corley, J.J. Hardon and B.J. Wood, Ch. 31, Elsevier, Amsterdam (1976), pp. 447–66.
46. P.D. Turner, 'Diseases of oil palm', in *Exotic Plant Quarantine Pests and Procedures for Introduction of Plant Materials*, ed. by K.G. Singh, ASEAN Plant Quarantine Centre and Training Institute, Serdang (1983), pp. 35–51.
47. J.J. Hardon and R.H.V. Corley, 'Pollination', in *Oil Palm Research*, ed. by R.H.V. Corley, J.J. Hardon and B.J. Wood, Ch. 21, Elsevier, Amsterdam (1976), pp. 299–305.
48. R.A. Syed, 'Studies on oil palm pollination by insects', *Bull. Ent. Res.*, **69**, 213–24 (1979).
49. R.A. Syed, 'Pollinating thrips of oil palm in West Malaysia', *Planter*, **57**, 62–81 (1981).
50. *Planter* Editorial, 'Too good to be true?', *Planter*, **57**, 59–61 (1981).
51. B.J. Wood, 'Note on insect pollination of oil palm in South and Central America', *Planter, Kuala Lumpur*, **59**, 167–170 (1983).
52. S.M. Kang and Zam bte. A. Karim, 'Quarantine aspects of the introduction into Malaysia of an oil palm insect pollinator', in *Plant Protection in the Tropics*, ed. by K.L. Heong, B.S. Lee, T.M. Lim, C.H. Teoh and Yusof Ibrahim, Malaysian Plant Protection Society, Kuala Lumpur (1982), pp. 615–26.
53. PORIM Editorial, *Proceedings of Symposium on Impact of the Pollinating Weevil on the Malaysian Oil Palm Industry*, 21–22 February 1984, Palm Oil Research Institute of Malaysia (PORIM), Kuala Lumpur, No. 8, pp. i–viii; 1–376 (1985).
54. B.J. Wood, S.G. Loong, I. Said, M.H. Lee and S.K. Quah, 'Mill recovery of palm oil from fresh fruit bunches (FFB) harvested to various "ripeness" standards', *Proceedings of Symposium on Impact of the Pollinating Weevil on the Malaysian Oil Palm Industry*, 21–22 February 1984 PORIM/MOPGC, Kuala Lumpur, pp. i–viii; 1–376 (1985).
55. S.Y. Foo, Mohd. Saleh bin Mohd. Tahir, L.P. Munusamy and J.E. Duckett, 'The use of aluminium poles in tall palm harvesting', *Planter*, **57**, 313–20 (1981).
56. R.H.V. Corley and B.S. Gray, 'Yield and yield components', in *Oil Palm Research*, ed. by R.H.V. Corley, J.J. Hardon and B.J. Wood, Ch. 6, Elsevier, Amsterdam (1976), pp. 77–86.
57. J.W.L. Bevan and B.S. Gray, *The Organisation and Control of Field Practice for Large-scale Oil Palm Plantings in Malaysia*, Incorporated Society of Planters, Kuala Lumpur (1969), 166 pp.
58. B.J. Wood, 'Research in relation to natural resource — Oil palm', *Planter, Kuala Lumpur*, **54**(628), 414–41 (1978).
59. Tan Sri Datuk and Anuwar bin Mahmud, 'Oil palm — Prospects and challenges towards the year 2000', Proceedings of Conference on *ASEAN Agriculture in the year 2000*, Agricultural Institute of Malaysia, Serdang, in the press.

3 Extraction of crude palm oil

J.H. Maycock

3.1	Bunch reception	29
3.2	Sterilizing	29
3.3	Stripping	32
3.4	Digestion	33
3.5	Oil extraction	34
3.6	Clarification and purification of crude palm oil	34
3.6.1	Gravity settling	35
3.6.2	Direct clarification system	36
3.7	Process control	37

This chapter deals with the process of extraction of crude palm oil from oil palm bunches. The flowchart of the process is shown in Figure 3.1; the separation of kernels and extraction of palm-kernel oil is not covered in this report.

3.1 Bunch reception

Bunches are transported to the mills by trucks or cages on a narrow-gauge railway system or by road with motor lorries or tractor trailers. The railway system, with its large capital outlay, has mainly given way to road transport. The adoption of road transport calls for the quick transfer of FFB (fresh fruit bunches) from the lorries and/or trailers to the sterilizer cages using bunch loading ramps. The first ramps were elevated platforms with 'hand' loading but present 'self-discharging' ramps are equipped with remotely operated gates. In an endeavour to produce special-quality oil some estates transport sterilizer cages to the field by lorry or tractor trailers, to keep handling and bruising of the fruit to a minimum.

3.2 Sterilizing

The objects of sterilizing can be briefly described:
1. Prevention of any further rise in FFA (free fatty acid) due to enzyme action by inactivation of the lipolytic enzymes.

29

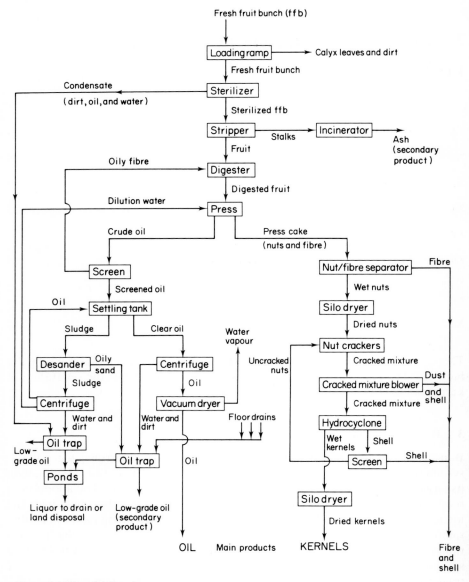

Figure 3.1 Material flowchart.

2. Facilitation of mechanical stripping. To loosen the fruit still attached to the bunch stalk sufficient heat must penetrate to the points of attachment of the fruits to the spikelets to bring about hydrolysis at these points.
3. Preparation of the fruit pericarp for subsequent processing.

4. Preconditioning of the nuts to minimize kernel breakage during both pressing and nut cracking.
5. Coagulation of the protein material and hydrolysis of the mucilaginous material present in the palm fruit.

The horizontal sterilizers now commonly used have internal rails for ease of movement of the open-topped perforated steel cages, mounted on wheeled chassis, which contain the FFB. In most mills cages are of 2.5 t capacity (older mills still have 1.5 t cages); the latest thinking is to go to even bigger cages — one mill in Malaysia has adopted 3.5 t cages. The sterilizers may be provided with a single door or with a door at each end. The latter arrangement is preferred at larger capacity mills as it simplifies the movement of the cages.

The maximum steam pressure used for sterilizing is usually 40 lbg/in² (the temperature of saturated steam at this pressure is 141.5 °C), taken directly from the boiler main via a reducing valve or, more commonly, from the exhaust of the steam-driven alternator set. The FFB must be allowed to heat thoroughly and become 'cooked'. With satisfactory sterilizing the temperature reached in the centre of the bunch stalk will be at least 100 °C, the time depending on the weight of the individual bunches.

An important factor in determining the length of the sterilizing cycle is the pressure within the vessel during the cooking period since each 10 °C rise in temperature reduces the cooking time by a factor of 2. Although the length of the sterilizing cycle could be reduced by increasing the operating pressure above 40 lbg/in² this is not advisable as the bleachability of the oil would be adversely affected. Another important factor affecting the sterilizing process is the presence of air in the vessel. A mixture of steam and air has a lower temperature than steam alone, so air removal is essential.

Originally, a single-peak sterilizing cycle was normally adopted — i.e. after the preliminary venting of the vessel the pressure was allowed to build up to 40 lbg/in², with a holding, or 'cooking', period of 60 minutes. At the end of this holding period the condensate and exhaust valves were opened and the pressure was allowed to drop to atmospheric. The main limitation of single-peak sterilizing is that air is liberated from the bunches during the holding period and cannot be removed from the vessel, so cooking is less efficient and takes longer. This was more obvious with the introduction of the weevil (see Chapter 2) on the Malaysian scene, producing larger and more compact bunches. Multiple-peak sterilizing cycles are now adopted if sufficient steam is available. The most usual form of this (triple-peak sterilization) involves raising the pressure initially to 30 lbg/in² and then exhausting to atmospheric pressure, raising the pressure again, this time to 35 lbg/in², and then exhausting to atmospheric pressure and finally raising the pressure to 40 lbg/in², at which pressure it is held for approximately 30–40 minutes.

The exact timing of a typical three-peak sterilizing cycle cannot be specified since this will depend on the size of the sterilizer and the steam supply arrangement; however, the following figures have been recorded at one mill:

Preliminary venting	2 min
Raise pressure to 30 lbg/in^2	6 min
Exhaust to atmospheric pressure	1 min
Venting	2 min
Raise pressure to 35 lbg/in^2	5 min
Exhaust to atmospheric pressure	2.25 min
Venting	2 min
Raise pressure to 40 lbg/in^2	3 min
Holding at 40 lbg/in^2	35 min
Exhaust to atmospheric pressure	3 min
Empty	5 min
Refill	5 min
Total cycle	71.25 min

After sterilization the cages are manoeuvred towards the main processing area and lifted by a hoist to the stripping section.

3.3 Stripping

The object of stripping, sometimes called 'threshing', is to separate the sterilized fruits together with the associated calyx leaves from the sterilized bunch stalks. Two main types of stripper have been developed: the beater arm and the rotary drum, which may be of the shaft or shaftless type. The beater was normally used for small capacity mills but beater arm strippers have been installed after the rotary drum type in mills of larger capacity to act as secondary strippers. With the beater stripper the bunches pass along a cradle made from a large number of curved bars. Each bar is separated from its neighbours by a fixed distance and the whole is made rigid by spacers and tie bars. Parallel to and beneath this horizontal cage is a rotating shaft carrying pairs of beater arms. These pairs of arms are spaced by the same distance separating the curved bars. By adjusting the height of the cage the tips of the beater arms are made to project somewhat between the curved bars. Each pair of beater arms is set at an angle to the adjacent pair so that, when the shaft rotates, not all the beater arm tips project at once but they do so in sequence. This ensures that any bunch of sterilized fruit placed on the cradle is subjected to one beating action for each revolution of the shaft by each beater arm along the section of the cradle which it occupies at any instant. This action knocks out the fruit together with the calyx leaves, and these fall between the curved bars into a screw conveyor, which takes them on their way to the digesters.

The rotary drum type of stripper consists of a long horizontal cylindrical drum which is rotated. The sterilized bunches are fed in continuously at one end and stalks pass out continuously at the other end. The cylindrical surface of the drum is made up of tee-bars running parallel to the axis of the cylinder and spaced far enough apart to permit the escape of the fruit and yet close enough to prevent stalks from passing out between them. Drum strippers are often built with a central shaft carrying spiders to which the cylindrical cage is attached and with the shaft itself supported on bearings. This is satisfactory for short drum strippers but for longer machines a shaftless form of construction is preferable and in this case rollers are used to support the drum, the cage being provided with two cast steel tyres which turn on the rollers. The rate of rotation of the drum is such as to ensure that the bunches of normal size are lifted by the centrifugal action, assisted by lifting bars fitted to the inside of the drum. Once the bunches have reached the top of the drum they then fall freely, passing approximately through the axis of the drum and striking the bottom with sufficient force to dislodge much of the fruit. The latter passes out between the tee-bars and falls into a screw conveyor which removes it. The partly threshed bunch is lifted up and falls down again; this action is repeated a number of times, leading to the removal of all the fruits as the stalk gradually works its way towards the end of the cage and finally drops out.

The maintenance of a regular feed of bunches to the stripper is desirable. This is facilitated in a modern mill by the use of a bunch feeder, usually in the form of a very strongly built variable-speed scraper conveyor onto which the cage of sterilized bunches is emptied by tipping it while it is still suspended from the hoist.

There are two discharges from the strippers — sterilized fruit and empty bunch stalks. The fruit goes to the digesters and the stalks are normally passed to an incinerator yielding an ash rich in potash which can be used as a fertilizer.

3.4 Digestion

After the bunches have been stripped, the sterilized fruit and accompanying calyx leaves must be reheated and the pericarp loosened from the nuts and prepared for pressing. This is carried out in steam-heated vessels provided with stirring arms known as digesters or kettles. Digesters commonly used are vertically arranged cylindrical vessels having rotating shafts to which are attached stirring arms. These arms stir and rub the fruit, loosening the pericarp from the nuts and at the same time breaking open as many of the oil cells as possible. The normal speed of the stirring arms is approximately 26 r/min.

The digester is kept full and as digested fruit is drawn off continuously, or intermittently, from near the bottom of the vessel so freshly stripped fruit is added at an equal rate. It is essential for good digestion that the level of fruit in the digester be kept as high as possible at all times. This ensures maximum holding time and maximum stirring effect per revolution. The latter depends on the pressure to which the fruit in the lower part of the digester is subjected.

The temperature of the digested fruit must be close to 100°C, which can be achieved by fitting the digester with a steam jacket or with live steam injection or a combination of both. With inadequate digestion there is a tendency for high oil losses in the press fibre.

3.5 Oil extraction

Although many systems, both wet and dry, have been used over the years to extract the crude oil from the digested fruit mash, it is common practice nowadays to use screw presses, especially when tenera palm fruits have to be processed. Screw presses consist essentially of a perforated cage in which runs a single or a double screw. The outlet end of the perforated cage is restricted by a cone, or cones, and it is this restriction of the discharge that creates a pressure in the cage and thus controls the amount of de-oiling of the digested fruit mash. Many types are available based on the following principles:
1. Two screws or worms of opposite hand running in opposite directions on the same centrally arranged shaft in a cylindrical cage.
2. Two screws or worms running side by side in a common perforated cage having the shape of a 'figure eight' in cross-section. The screws can be arranged horizontally or vertically but with the vertical arrangement a feed screw is normally provided.
3. A single screw or worm running in a perforated cage. With this design, it is normal to provide a feed screw.

The extent of de-oiling of the press fibre occurring in a screw press depends on the design and mechanical condition of the press, the composition of the fruit and the way the press is operated. To ensure maximum de-oiling and minimum nut breakage the fruit should be properly digested and at a temperature close to 100 °C when fed to the press. The cone must be positioned close enough to the cake outlet at the end of the press cage to ensure that the cake is put under sufficient pressure without nut breakage becoming too serious or the driving motor becoming overloaded. In practice with electric-motor-driven presses the cone is positioned, as far as is possible, to ensure that the motor current is always up to a value that will not overload the motor or press and experience shows gives suitable results. The way in which the cone position is changed depends on the press construction. Movement of the cone by a manually operated hydraulic system allows the operator to make rapid cone adjustments guided by the ammeter readings. Alternatively the cone position is varied by hydraulic pressure, which is varied automatically whenever the motor current exceeds a preset maximum value or falls below a preset minimum value. Thus if the current rises to a certain value the hydraulic pressure falls and the cone moves out a little, so lowering the load. Similarly, if the current falls below a preset level the hydraulic pressure is raised, thereby moving the cone in slightly and making the load increase.

Crude palm oil liquor and a matte of oily fibre and nuts are discharged from the screw presses.

3.6 Clarification and purification of the crude palm oil

Crude oil extracted from palm fruit by pressing contains varying amounts of water together with impurities consisting of vegetable matter, some in the form of insoluble solids and some dissolved in the water.

An average composition of crude palm oil obtained from a screw press is as follows:

	Oil (%)	Water (%)	(Non-oily solids) (NOS) (%)
Average	66	24	10
Range	40–75	10–40	6–25

Considerable deviations from the average figure are possible, as given above, depending on the type and condition of the press and also on the amount of water dilution before or during pressing.

To give a clear stable product of acceptable appearance water and impurities must be removed. This is carried out in the clarification section of the mill. When the dehydrated, clear palm oil leaves the clarification section for the storage tanks it still contains a very small amount of soluble solids known as 'gums'. If the palm oil is not properly dried these gums will hydrate slowly over a period of time and become insoluble. This causes the deposition of 'foots' in storage tanks.

The water present in the crude palm oil can largely be removed by settling or centrifuging since most of it is free or undissolved. A small proportion, however, is dissolved in the oil and can only be removed by evaporation in a dehydrator with or without the use of vacuum. In practice not all the dissolved water is removed during dehydration as this would make the oil more susceptible to oxidation. The dissolved moisture content is in fact reduced from approximately 0.25 per cent to approximately 0.1 per cent, which is sufficient not only to prevent the deposition of foots but also, perhaps more importantly, to reduce the rate of hydrolysis so that the oil is practically stable during storage from the development of more free fatty acid.

Clarification systems normally used for crude palm oil are 'gravity settling' or the 'direct process', this latter system involving the use of decanters without any previous settling.

3.6.1 Gravity settling

With gravity settling (Figure 3.2) it is first necessary to dilute the crude oil with hot water to reduce its viscosity. The diluted crude oil is then screened to remove any coarse fibrous material which is returned to the digesters. The screened crude oil is then heated to appoximately 90–95 °C and pumped to the continuous settling tank. The retention of the crude liquor in this tank enables the oil to rise to the surface and overflow continuously into a reception tank. This settled oil is further

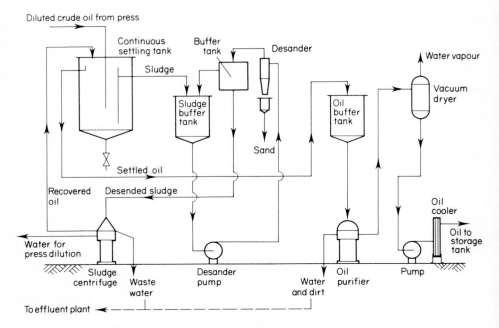

Figure 3.2 Diagrammatic arrangement of clarification plant with continuous settling tank.

purified by centrifuging which reduces the dirt content to 0.01 per cent or less. Finally, centrifuged oil is passed to a vacuum dryer to give a product with a moisture content of approximately 0.1 per cent. In some mills the oil is cooled to approximately 50 °C before being pumped to the storage tanks as this helps to lower the rate of oxidation of the oil.

The underflow from the continuous settling tank, generally referred to as 'sludge', contains some oil which is recovered by centrifuging. It is advisable to remove as much as possible of the sand from the sludge before passing it to the centrifuge because it causes rapid and severe wear to the bowl. The normal method of desanding is to use a specially designed hydrocyclone. The oil recovered from the underflow after centrifuging still contains some water and dirt and is therefore returned to the continuous settling tank for further treatment. The waste water from the sludge centrifuge contains a little oil which is very difficult, and uneconomical, to remove and is thus discharged to the effluent treatment plant.

3.6.2 Direct clarification system

In this system (Figure 3.3) undiluted crude oil from the screw presses is treated without the preliminary removal of some of the oil by settling. Although originally nozzle-type centrifuges were used for the first stage of centrifuging, for this sytem

decanters are now universally adopted. Again it is important that the crude palm oil is desanded before being passed to the decanters. From the decanter three fractions are obtained:
1. Oil containing 2.5% water and 0.5% non-oily solids
2. Water phase containing 1.5% oil and 11% non-oily solids
3. Solid or cake containing 2.5% oil and 23% non-oily solids
The water phase and the cake have a very low oil content, which is difficult and uneconomical to remove or reclaim. The water phase is therefore discharged to the effluent treatment plant. The cake is dried to provide animal feed or fertilizer.

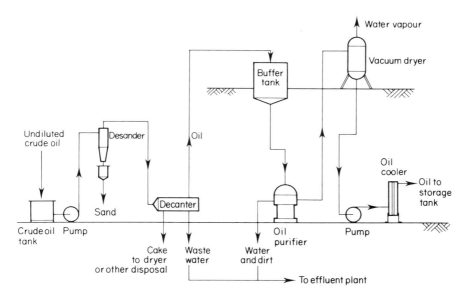

Figure 3.3 Diagrammatic arrangement of clarification plant with no settling using decanters (three phase).

The oil recovered from the decanter is passed to a centrifuge and then to a vacuum dryer as in the gravity settling system. The final oil is pumped from the clarification plant to storage tanks to await despatch.

3.7 Process control

This subject demands a chapter of its own but mention will now be made of the samples that should be taken and the tests that should be made to give a full control of the process, including not only a knowledge of the quality of the oil produced but also the extraction efficiencies.

Sample	Tests (quote as %)
1. Oil to storage	Free fatty acid Moisture Dirt Peroxide value
2. Oil as despatched	Free fatty acid Moisture Dirt Peroxide value
3. Waste water ex sludge centrifuges	Water Oil Dry non-oily solids
4. Bunch stalks	Fruit Water Oil Non-oily solids

4 Refining and fractionation of palm oil

F.V.K. Young

4.1	**The composition of palm oil**	39
4.2	**The quality of crude palm oil**	45
4.3	**Refining**	45
4.3.1	Introduction	45
4.3.2	Storage and handling of crude oil	46
4.3.3	Chemical refining	48
	(a) Gum conditioning	48
	(b) Neutralization, washing, drying	48
	(c) Bleaching	51
4.3.4	Physical refining pretreatment	55
4.3.5	Deodorization and distillation	57
	(a) Introduction	57
	(b) Process description	57
4.3.6	Refining after transportation	61
4.3.7	Effect of refining on oil composition	61
4.4	**Fractionation**	62
4.4.1	General considerations	62
4.4.2	Process description	64
	(a) Dry fractionation	65
	(b) Detergent fractionation	65
4.4.3	Products	68
4.5	**References**	69

4.1 The composition of palm oil

When considering the processes of refining and fractionation in relation to any edible oil, it is necessary first to examine the characteristics and composition of the oil. With this information it is possible to decide on the refining methods which can be used, on the practicality and process conditions for fractionation or other modification processes and on the manner in which the oil can be used in end products.

The oil palm tree is to be found mainly within a belt of $\pm 10°$ latitude around the equator. The vast bulk of the oil produced stems from the West African tree, *Elaeis guineensis*. In spite of differences in hybridization, climate and soil the oil obtained from the original stock is basically similar whether it is grown in Africa or South-East Asia (Tables 4.1 and 4.2). The South American oil palm, *E. oleifera*,

40   F.V.K. Young

Table 4.1 Characteristics of palm oil

	Codex[1]	Zaire[2]	Malaysia[3] Range	Mean value
Relative density, 50 °C	0.891–0.899	0.891	0.8919–0.8932	0.8927
Refractive index n_D, 50 °C	1.449–1.455	1.455	1.4546–1.4560	1.4553
Flash point (°C)	—	277	—	—
Melting range (°C)	—	27–50	32.3–39.0	36.0
Saponification value (mg of KOH/g of oil)	190–209	199	190.1–201.7	195.7
Iodine value (Wijs)	50–55	53.5	51.0–55.3	53.3

Table 4.2 Percentage fatty acid composition of palm oil

Fatty acid	Codex[1]	Zaire plantation[2]	Malaysia PORIM mean[3]	Colombia *Elaeis oleifera*[4]
12:0	<1.2	0.12	0.2	—
14:0	0.5–5.9	1.02	1.1	Trace
16:0	32–59	45.5	44.0	22.9
16:1	<0.6	0.10	0.1	1.3
18:0	1.5–8.0	5.90	4.5	1.0
18:1	27–52	34.6	39.2	54.8
18:2	5.0–14	11.8	10.1	20.0
18:3	<1.5	0.29	0.4	—
20:0	<1.0	0.36	0.4	—
Iodine value (Wijs)	—	53.5 (calc.)	53.3	78.5

is noticeably different in the fatty acid composition of the oil (Table 4.2) and, because of the higher content of linoleic acid, is more desirable. However, the yield of the South American tree is substantially lower than that of trees from the *E. guineensis* stock and therefore it is not at present grown commercially.

The refractive index of palm oil reflects its lower iodine value and higher melting point compared with liquid oils such as soyabean. It can be seen from Table 4.4 that the oil contains sufficient solid fat at its melting point of about 36 °C for it to require heating in order to keep it liquid in most parts of the world, so involving a significant risk of quality deterioration. The melting range for Zaire oil (Table 4.1) is extreme and includes oils which during extraction or when sampled have been subject to varying degrees of fractionation. The melting range for commercial oils of African origin is similar to that given for Malaysian oils.

The individuality of palm oil stems primarily from its high content of palmitic acid which, at about 45 per cent in present commercial oils, is almost twice the quantity found in other common oils rich in palmitic acid. This feature, together with the other fatty acids present in significant quantities, stearic, oleic and linoleic, is responsible for the triglyceride composition shown in Table 4.3 and the consequent solid fat content (SFC) curve of the oil in Table 4.4. As a result of its SFC curve, palm oil is widely used either on its own as a frying oil or in blends with

Table 4.3 Triglyceride composition of Zaire palm oil (mol %)[a]

Zero double bond		One double bond		Two double bonds		Three double bonds		Four or more double bonds	
Glyceride	Amount	Glyceride	Amount	Glyceride	Amount	Glyceride	Amount	Glyceride	Amount
PMP[b]	0.3	MOP	1.2	MOO	0.6	OOO	2.7	OOL	1.8
MPP	0.2	POP	24.1	POO	18.9	MOL	0.2	OLO	1.4
PPP	4.3	POS	7.0	SOO	2.8	POL	4.0	PLL	2.2
PPS	1.1	SOS	0.5	OPO	1.0	SOL	0.6	LPL	0.4
PSP	0.4	PPO	3.6	PPL	0.4	MLO	0.3	SOLe	
		SPO	0.5	MLP	0.5	PLO	4.5	SLeO }	0.7
				PLP	7.8	SLO	0.6	OSLe	
				PLS	2.3	OPL	0.3		
Others	0.3	Others	1.2	Others	0.7	Others	0.1	Others	0.5
	6.6		38.1		35.0		13.3		7.0

Percentage by weight of total fractions

[a] The composition was computed by Loncin et al.[2] from the data of Jurriens et al.[5].
[b] M = myristic, P = palmitic, S = stearic, O = oleic, L = linoleic, Le = linolenic.

Table 4.4 Solid fat content (n.m.r.^a) of Malaysian palm oil (%)[3]

Temperature (°C)	Mean values, PORIM samples		
	Crude	Neutralized	Refined
5	59.9	64.9	62.2
10	48.5	54.2	50.3
15	33.0	38.2	35.2
20	21.8	26.1	23.2
25	13.4	15.2	13.7
30	9.3	9.8	8.5
35	6.6	6.7	5.8
40	4.2	4.3	3.5

^aNuclear magnetic resonance spectroscopy.

other oils and fats for a variety of margarines, shortenings and specialist products. The use of the fractionation process in taking advantage of the triglyceride composition is described later.

The minor components of an oil consist of free fatty acids, mono- and diglycerides, the unsaponifiable fraction and impurities such as trace metals and oxidation products. They are very important in assessing the quality of the crude oil as they affect the quality of the end product and the efficiency of the refining and fractionation processes.

Free fatty acids are formed as a result of hydrolysis of the triglycerides. The hydrolysis reaction is effected by enzymatic action in the fruit and by extraction of the oil from the fruit, and will continue under adverse conditions in the crude oil. Crude palm oil normally contains between 2.5 and 5 per cent of free fatty acids when it reaches the refinery but values up to 10 per cent can be found in oils which have been badly handled.

The monoglyceride content of palm oil is normally less than 1 per cent. However, Koslowski[6] reported finding 5.5 per cent of monoglyceride in an oil of FFA 9%. The subject of diglycerides is more complex since they are intermediates in the synthesis of triglycerides *in vivo* and therefore their presence in palm oil is not solely due to hydrolysis. Jacobsberg and Oh[7] found a diglyceride content of 5.66% in a crude oil of FFA 0.3% which was extracted in the laboratory from fresh unbruised fruit. A second lot of the same fruit was mechanically bruised before extraction and yielded an oil with FFA 5.4% and a diglyceride content of 7.6%. In commercial oils of FFA between 3 and 5% the diglyceride content is usually between 5 and 8% but further hydrolysis which results in higher FFAs can produce diglyceride contents over 9%. These higher levels are of significance in the fractionation of the oil.

According to the Codex International Standard[1] the unsaponifiable fraction of palm oil amounts to less than 1.2 per cent of the total crude oil and the mean value for crude oils is about 0.5 per cent. As shown in Table 4.5 it is

Table 4.5 Unsaponifiable composition of palm oil

Component	Zaire[2]		Malaysia[8]	
	%	mg/kg of oil	%	mg/kg of oil
Carotenoids				
α-Carotene	36.2		29	
β-Carotene	54.4		62	
γ-Carotene	3.3	500–700	4	500–700
Lycopene	3.8		2	
Xanthophylls	2.2		3	
Tocopherols/tocotrienols				
α-Tocopherol	35		20	
γ-Tocopherol	35		—	
δ-Tocopherol	10		—	
ε + η-Tocopherol	10	500–800	—	ca. 800
α-Tocotrienol	—		25	
γ-Tocotrienol	—		45	
δ-Tocotrienol	—		10	
Sterols				
Cholesterol	4		4.1[9]	
Campesterol	21		22.8	
Stigmasterol	21	ca. 300	11.3	326–627[9]
β-Sitosterol	63		57.5	
Total Alcohols				
Triterpenic alcohols	80		—	
Aliphatic alcohols	20	ca. 800	—	
Phosphatides				
Phosphatidylcholine	—		36[10]	
Phosphatidylethanolamine	—		24	
Phosphatidylinositol	—	500–1000	22	20–80[10][a]
Phosphatidylglycerol	—		9	
Diphosphatidylglycerol	—		4	
Phosphatidic Acid	—		3	

[a]Expressed as phosphorus

composed of carotenoid pigments, tocopherols and tocotrienols, sterols, alcohols and phosphatides. The groups of most importance to the refiner are the carotenoids, tocopherols and tocotrienols and the phosphatides. The unsaponifiable fraction of palm oil has been reviewed by Goh *et al.*[11].

The carotenoids are the pigments responsible for the red colour of crude palm oil. They are thermally decomposed during the refining process. Figure 4.1 shows the degree and rate of decomposition of β-carotene at various temperatures[2]. However, if during extraction, shipment and storage the oil is mishandled, particularly by overheating in the presence of oxygen and traces of iron, condensation products of the carotenoids are formed. The latter products are responsible for a brown colour in the oil which is not removed by heat treatment and is difficult to remove by adsorption bleaching. In such cases the carotene content of the crude oil will be found to be below 450 ppm.

As well as constituting the family of E vitamins the tocopherols and trienols are natural antioxidants when present at less than 2000 ppm. Above that figure there

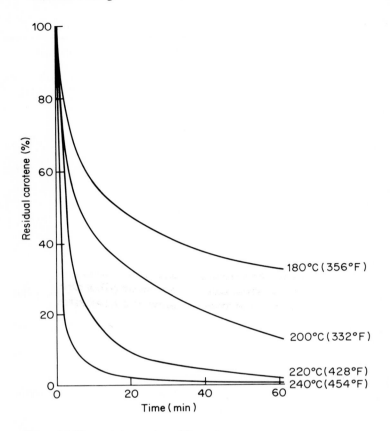

Figure 4.1 Thermal destruction of β-carotene.

is evidence for prooxidant activity. Palm oil contains less than 1000 ppm and it is therefore important that as far as possible they should be retained in the oil during the various process steps undergone by the oil.

The phosphorus compounds in palm oil have been stated by Goh *et al.*[10, 12, 13] to be comprised of inorganic phosphates and phospholipids. Phospholipids are better known to the refiner as phosphatides and are frequently referred to, together with small quantities of glucosides, carbohydrates and pectins, as 'gums', which have adverse effects on product quality and yield of refined oil. Phosphatides in edible oils are normally calculated from the determination of total phosphorus and the use of a factor relating the molecular weight of phosphorus to the mean molecular weight of the phosphatides in the oil. The Zaire figure (Table 4.5) was so calculated. The total phosphorus figure of Goh *et al.*[10] is 20 ppm which, for the sake of comparison and by using their factor of 24, can be expressed as a phosphatide content of 480 mg/kg.

4.2 The quality of crude palm oil

The quality of the crude oil is determined by the levels of impurities in the oil. These fall into two groups according to their effect on the oil:
1. Hydrolytic, e.g. moisture, insoluble impurities, free fatty acids, partial glycerides, enzymes.
2. Oxidative, e.g. oxidation products, trace metals, pigments.

Phosphatides are emulsifiers and so hinder the separation of oil and water phases in the chemical refining process. If they are not removed prior to physical refining they char at the high temperatures employed, giving the oil a brown colour. The phosphatides are broadly separated into hydratable and non-hydratable types. As the name implies, hydratable phosphatides can be removed by treatment with water, while the non-hydratable compounds, which are salts or coordination compounds of calcium and magnesium primarily with phosphatidic acid, can only be rendered insoluble in the oil by the use of chemical reagents, the most commonly used being phosphoric acid (H_3PO_4). Carbohydrates, pectins and glucosides increase the viscosity of the oil with a consequent detrimental effect on refining and fractionation efficiency.

As far as the quality of the crude oil is concerned the oil palm fruit mesocarp contains three important enzyme systems: the lipases responsible for hydrolytic splitting of the triglycerides, the lipoxygenase enzymes which catalyse oxidation of the fatty acid chains and the phosphorylases which cause an increase in the phosphatide level. When the fruit is bruised, rupture of the oil cell walls allows access of the enzymes to the oil with the above results.

Palm oil is extracted by a process which involves considerable contact of the oil with water at temperatures around 90 °C. Although part of the plant is constructed from stainless steel, the greater part is of mild steel, so the risk of iron contamination and iron-catalysed oxidation of the oil is considerable. Copper is a strong prooxidant and contact between the oil and copper in any form should be prevented.

The quality of the oil is judged by the refiner mainly using the parameters shown in Table 4.6. The carotene content is included because it is a good indicator of the bleachability of the oil. Low levels of carotene suggest that the oil will be difficult to bleach because of the formation of the coloured condensation products referred to earlier.

4.3 Refining

4.3.1 Introduction

Purification or refining processes are needed to reduce as far as possible those contaminants of the crude oil which will adversely affect the quality of the end product and the efficient operation of the modification processes by fractionation, hydrogenation and interesterification. At the same time the refining treatment

Table 4.6 Crude palm oil quality guidelines

Source	Codex[1]	Commercially available[14] qualities		Author's experience
		SQ[a]	Normal	
Moisture (%)	0.2 max.	—	—	0.15–0.20
Insoluble impurities (%)	0.05 max.	—	—	<0.05
Diglycerides (%)	—	±4	Above SQ	±5
Monoglycerides (%)	—	±0.1	±0.2	0.2–0.3
Free fatty acids, palmitic (%)	Acid value max. 10	<2	5 max.	±4.5
Phosphorus (mg/kg)	—	<5	±20	15–35
Tocopherols (mg/kg)	—	±800	Below SQ	—
Carotene (mg/kg)	500 min., 2000 max.	±550	±550	400–550
Peroxide value (meq/kg)	10 max.	—	—	±10
Anisidine value	—	—	—	±5
Totox value[b]	—	±5	Above SQ	±25
Iron (mg/kg)	5 max.	±3	Above SQ	±4
Copper (mg/kg)	0.4 max.	±0.02	±0.05	±0.02

[a]Special quality.
[b]Totox value (TV) = total oxidation value = 2 × peroxide value + anisidine value.

should retain as much as possible of the tocopherols and tocotrienols because of their antioxidant effect on the product.

Two methods are in use: these are termed 'physical' and 'chemical' from the means by which the free fatty acids are removed from the oil. The fatty acids are distilled off in the physical process, and in the chemical process are neutralized using an alkaline reagent, thus forming soaps that are removed from the oil by phase separation. The process steps involved in the two methods are depicted in Figure 4.2 and the principal impurities removed are shown in Table 4.7.

Of these methods the physical process is the more efficient, in terms of yield of product and utilities consumed, but the chemical process is more flexible in that it can provide an acceptable product from poorer quality raw material. The choice of method is therefore dependent on the quality of the product required by the market and the quality of the crude oil. Data in Table 4.6 can be examined in this context. If the desired product quality is to be of maximum 3.0 red units Lovibond in a 5.25 in cell and minimum 50 h oxidative stability by the AOM test, this should be obtained from both of the oils quoted by Jacobsberg[14] by physical refining. The same quality specification could be obtained by chemical refining (from the oil data given by the author) but probably not by physical refining.

The physical process also presents a much simplified effluent treatment problem because no soap is produced. As a result of these factors by far the major part of the world's palm oil production is today refined by the physical method.

4.3.2 Storage and handling of crude oil

The storage and handling of the crude oil, of the utmost importance in obtaining a

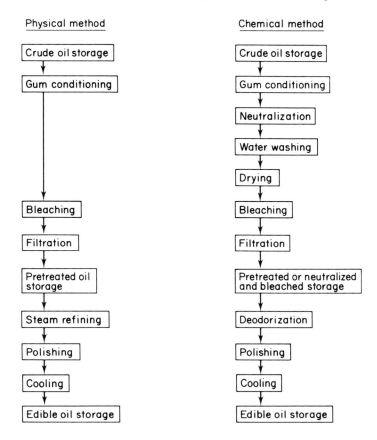

Figure 4.2 Oil refining stages.

Table 4.7 Refining — unit processes

Stage	Principal impurities reduced or removed
Degumming	Phospholipids, trace metals, pigments, carbohydrates, protein
Neutralization	Fatty acids, phospholipids, pigments, trace metals, sulphur, oil insolubles, water solubles
Washing	Soap
Drying	Water
Bleaching	Pigments, oxidation products, trace metals, sulphur, traces of soap
Filtration	Spent bleaching earth
Deodorization	Fatty acids, monoglycerides, oxidation products, pigment decomposition products
Physical refining	Fatty acids, monoglycerides, oxidation products, pigment decomposition products
Polishing	Removal of trace oil insolubles

48   F.V.K. Young

high yield of a high-quality refined product, is the subject of a booklet by the Palm Oil Institute of Malaysia, PORIM[15]. Factors of importance for oil quality are:
1. prevention of contamination between different oil types,
2. minimizing of free fatty acid increase, of oxidation and of colour fixation,
3. prevention of unintentional fractionation.

Moisture in the oil causes increases of the free fatty acid content and also gives rise to rust, which causes higher oxidation levels and colour fixation due to increased iron contamination. Overheating has the same effect. Conversely, if the oil is too cold, high melting point triglycerides will crystallize and cause variability in the composition of the oil being withdrawn from the tank.

4.3.3 Chemical refining

(a) Gum conditioning Gum conditioning is the step in the chemical refining process which precedes the neutralization of the free fatty acids by alkali. Its purpose is primarily to convert the non-hydratable phosphatides into a form in which they can be removed from the oil during soapstock separation without causing losses due to emulsification.

The chemical most frequently used is phosphoric acid in concentrated (75–85 per cent) form. This decomposes the non-hydratable phosphatides forming a dark brown sludge. It also decomposes iron salts. Because of the decomposition of calcium and magnesium salts the soapstock is more easily separated and consequently higher yields and quality of the refined oil is obtained.

Gum conditioning is employed in batch, semicontinuous and continuous processing. The quantity of phosphoric acid used is based on the phosphorus content of the crude oil and is usually 0.05–0.1 per cent of the oil weight. The acid must be intimately mixed with the oil. In the United States, using continuous plant, acid is added to the oil at 45–50 °C and, after a reaction time of up to four hours, the caustic soda neutralizing agent is added prior to heating to the centrifugal separation temperature of 80 °C. In most other parts of the world the oil is heated to 85–90 °C, acid added with intense agitation and, following a residence time of up to 5 minutes, the oil is treated with caustic soda. The residence time permits agglomeration of the denatured phosphatide particles, thus assisting separation.

(b) Neutralization, washing, drying For palm oil the batch neutralization process[16] has been superseded by continuous and semicontinuous lines because of improved oil quality and yield and the higher degree of control possible with these units. Therefore in this section reference will be restricted to continuous processes.

(i) General considerations The objectives of the neutralization step are removal of the free fatty acids, dirt and denatured phosphatides together with

pigments and other saponifiable impurities. These objectives must be accomplished with minimum neutral oil losses from emulsification, occlusion in the soapstock and saponification. The process variables include:

1. the type of alkali used, e.g. sodium hydroxide (caustic soda), sodium carbonate (soda ash), ammonium hydroxide, sodium silicate;
2. the strength of the alkali solution;
3. the excess used over the stoichiometrically required amount;
4. the temperature of addition;
5. the degree of agitation and
6. the oil flowrate through the plant.

The most widely used alkali is sodium hydroxide because of its thorough cleansing action on the oil. A strong caustic soda, for instance 20°Bé, is effective for the removal of impurities, especially gums and pigments, but saponification losses are higher and the soap is also thicker than when using weak lye (6°Bé), with a consequently higher loss due to oil occlusion in the soapstock.

On the other hand, emulsion losses are higher when using dilute solutions. Similarly, a large excess (30–50 per cent) has a high purifying action but a greater tendency to saponification. Emulsions are more likely to be encountered with low excess lye usage (5–15 per cent).

A long contact time, high mixing intensity and high temperature are responsible for high saponification losses but a long contact time helps to remove pigments and high temperatures improve phase separation.

A double neutralization technique is often used for the treatment of poor quality oils. In the first stage a strong (20°Bé) caustic soda is used and, after soap separation, the oil is given a second lye treatment using strengths between 6° and 20°Bé, depending on the quality of the oil.

The modern, fully continuous centrifugal refining line allows for many variations of throughput, mixing speed, contact time and temperature so that the influence mentioned can be made use of to optimize quality and yield for a wide variety of raw material quality.

After centrifugal separation of the soapstock from the oil, the oil contains between 500 and 1000 ppm of soap, which is removed primarily by washing with water. The water for this purpose and for lye make-up should be soft; steam condensate is suitable and should always be hotter than the oil (preferably over 90°C). The quantity of water used is between 10 and 20 per cent of oil flow.

Where two washing stages are used in centrifugal refining, plant manufacturers offer a countercurrent washing system in which the spent wash water from the final centrifuge is used again to wash the oil in the first washing step. Thorough washing is needed. Soap is adsorbed on bleaching earth, thus reducing the earth's surface availability for the removal of other impurities. The efficiency of water washing depends mainly on the removal of phosphatides with the soapstock, but with a well-refined oil the soap content will be less than 50 ppm after two washing stages. To eliminate soap from the washed oil, citric or phosphoric acid may be added to the final wash water or an aqueous citric acid solution can be added prior to drying the oil.

The soapstock and water washes are combined and 'split' using mineral acid, usually sulphuric acid. The oil phase, known as 'acid oil', is settled, and in some cases washed, to achieve the required specification and is sold as a byproduct. The aqueous phase is run to effluent.

(ii) Plant Two basic systems are in use for the continuous centrifugal refining of edible oils[17, 18]. They differ in the contact time and temperature of initial contact between the oil and the caustic soda solution. In the United States the lye is added to the oil at low temperature (palm oil 50 °C) and the oil–lye mixing time is between 5 and 15 minutes. At the end of this time the mixture is heated to 80 °C to coagulate the soap and is then immediately separated by centrifuge. This is known as the 'long mix' process. In Europe and many other parts of the world the 'short mix' process is used (Figure 4.3). The oil is heated to 90 °C, treated with phosphoric acid for gum conditioning and then mixed with lye for 1–15 seconds before separation. Re-refining, if required, washing and drying are similar for the two systems.

In its simplest form the continuous refining line incorporates two solid-bowl disc centrifuges. The first is used for soapstock separation and, to aid the removal of soapstock, is equipped for flushing the bowl periphery with water. The second centrifuge is used for water washing. This type of line may be used for the treatment of clean, high-quality oils that can be refined easily. However, dirt in the crude oil and denatured phosphatide sludge from gum conditioning hamper efficient line operation by accumulating on the surface of solid bowls. It is therefore a frequent practice for soapstock separation to use a split-bowl centrifuge with programmed periodic discharge of the dirt.

In most parts of the world the variability of oil quality requires that the continuous line should contain three separation stages as depicted in Figure 4.3. For reasons stated above, the first centrifuge, used for soapstock separation, is of the split-bowl, automatic discharge type. The second stage is used either for a second alkali refine or for a first wash. The plant is therefore equipped at this stage for the addition to the oil of either lye or water, and the centrifuge has a solid bowl with the possibility of using bowl-flush water. The third separation stage is used only for water washing and therefore does not require the bowl-flush water system. After washing, the oil may be mixed with citric acid solution to decompose remaining traces of soap before entering the drier, where water is removed from the oil by cascading or being sprayed under vacuum.

Alkali refining is also carried out in a semicontinuous manner using the Zenith Process[16], where small batches of oil are measured and treated with phosphoric acid in a pretreatment unit. From this unit the oil is pumped continuously via a centrifuge which removes dirt and denatured phosphatides to one of two neutralizer vessels that are used alternately. The oil enters the neutralizers which contain a dilute (0.3–0.4 N) caustic soda solution via a dispersing system and rises through the lye in streams of small droplets. The free fatty acids are neutralized and the soap thus formed is dissolved in the very weak solution. The oil settles out on top of the lye before passing continuously to a citric acid dosing and bleaching unit followed by filtration. The oil leaving the neutralizer contains less than 100 ppm of soap and therefore does not need to be washed. Traces of soap are decomposed by the citric acid before the oil is bleached. The removal of dirt before neutralization and the particularly mild neutralization conditions result in high yields being obtained in the refining of palm oil.

(c) Bleaching

(i) General considerations For many years the bleaching process, which consists of contacting the oil with an adsorbent clay or carbon, was considered purely from the standpoint of the removal of pigments from the oil. These pigments are primarily carotenoids, chlorophyll, gossypol and related compounds, and the products of degradation and condensation reactions that occur during the handling, storage and treatment of the raw materials and the extracted oils. It was later realized that activated adsorbents in particular are responsible for removing, at least partially, other impurities such as soap, trace metals and phosphatides. Oxidation levels are also reduced by breakdown of the hydroperoxide primary oxidation product on the adsorbent surface, followed by adsorption of the carbonyl compounds which are the secondary oxidation products. Due to the decolourization which occurs during modern high-temperature deodorization, these latter properties of the adsorbent have become the more important.

Segers[19] described the properties of activated bleaching earth as:
1. adsorption capacity, dependent on the surface area, pore size and pore volume;
2. catalytic activity, dependent on the degree of acid treatment and temperature of bleaching;
3. ion exchange activity, dependent on acid activation for the replacement of aluminium ions by which protons that give the earth certain metal-binding properties.

Bleaching must be carried out in the absence of oxygen, since the activated clay may act as a catalyst for oxidation if oxygen is present. Without this catalytic possibility oxidation would in any case occur in the presence of oxygen at the elevated temperature used. The process is therefore carried out under a steam or nitrogen blanket or, more usually on plant scale, under vacuum.

While vacuum has a beneficial effect in the removal of oxygen its effect in removing water from the system is more open to question. Brimberg[20], in bleaching rapeseed oil, increased the initial moisture content of the system from 0.056 to 0.477 per cent with a consequent increase in both the rate of reaction and overall removal of carotene and chlorophyll. However, in a later work on palm oil[21] the best results on carotene adsorption were obtained with 0.1 per cent

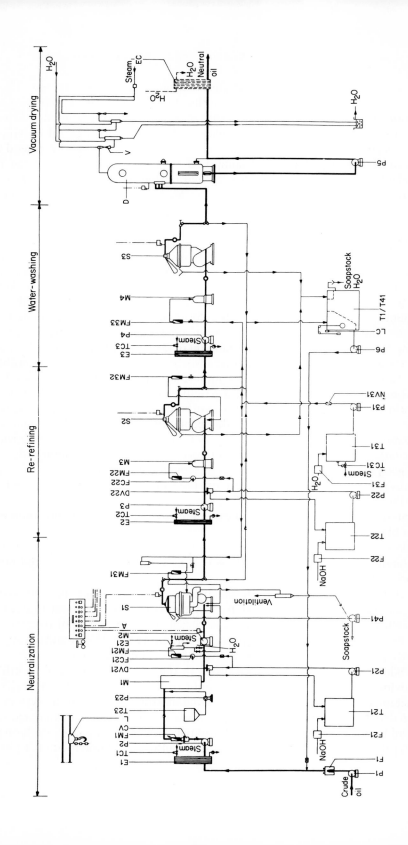

Centrifugal separators:

S1   Hermetic self cleaning separator
S2, S3   Hermetic separator

Pumps:

P1   Crude oil pump
P2, P3, P4 Oil feed pump
P5   Refined oil pump
P6   Slop oil pump
P21, P22   Lye pump
P23   Phosphoric acid pump
P31   Water pump
P41   Soapstock pump

Heat exchanges:

E1, E2, E3 Oil heater
EC   Oil cooler
E21   Lye heater

Mixers:

M1   Degumming mixer
M2   Neutralizing mixer
M3   Re-refining mixer
M4   Water-washing mixer

Strainers:

F1   Crude oil strainer
F21, F22 Lye strainer
F31   Water strainer

Controllers:

CV   Constant oil-flow controller
DV21, DV22 Lye-dosing ratio controller
FC21, FC22 Lye-flow control valve
LC   Oil level controller
TC1, TC2, TC3 Oil temperature controller
TC 31   Water temperature controller
NV 31   Non-return valve for water

Flow meters:

FM1   Oil flow meter
FM21, FM22 Lye flow meter
FM31, FM33 ⎫
FM33   ⎬ Water flow meter

Tanks:

T21, T22 Lye tank
T23   Phosphoric acid tank
T31   Water tank
T1/T41   Slop tank/catch basin

Vacuum dryer:

D   Vacuum dryer
V   Steam ejector

Auxiliary equipment:

A   Alarm system
L   Lifting tackle

Figure 4.3 Continuous caustic soda refining plant. (Reproduced by permission of Alfa Laval AB).

moisture in the system. Zschau[22] has reported that, although the colour of palm oil after bleaching with earth and water was not as good as with activated earth alone, it was better after the subsequent deodorization step. Other process parameters of importance are the temperature and the earth–oil contact time.

Process temperatures between 80 and 180 °C have been advocated for the process. Adsorption effects are usually improved by the use of high temperatures but in the treatment of low-quality oils the reverse can be true[22]. Because of damage to the oil caused by the ingress of air in industrial plant the temperatures employed for palm oil are normally between 90 and 120 °C. Kinetic studies reported by Brimberg[20, 21] indicate that the equilibrium stage of the adsorption of pigments is substantially reached after 4–40 min for activated earths, depending on the concentration and activity of the earth and the temperature of the process. Industrially, contact times of 10–30 min prior to filtration are employed but, in a patented process, Mag[23] claims that efficient adsorption can be achieved in less than three minutes contact time.

The quantity of activated earth required depends on quality of the oil, activity of the earth and the process conditions, and is normally between 0.5 and 2 per cent of the weight of the oil.

(ii) Plant Bleaching is carried out in batch, semicontinuous and fully continuous plant. In the batch process the bleaching step is often carried out in the same vessel as is used for neutralization, washing and drying. The vessel is referred to as a neutralizer-bleacher. Alternatively, a separate vacuum bleaching vessel is used. The batch operation is not as efficient as the other two methods, involves long earth–oil contact time during the filtration step and, because the batch is usually open to atmosphere during filtration, the oil is more prone to oxidation damage.

In a semicontinuous plant (Figure 4.4) the oil is pumped via a heat exchanger and deaeration/drier unit into the main vessel, which contains four compartments referred to as trays, all of which are maintained under a vacuum of 25–50 mmHg absolute. Oil is continuously pumped from the bottom tray to the filters and final cooling heat exchanger. The plant is automatically controlled using float switches, so that discrete small batches of oil are passed through the trays in the vessel. Bleaching earth and, if desired, citric acid is added to the deaerated oil in the first tray. The oil receives a defined minimum bleaching time in the second and third trays before passing to the fourth for filtration. If desired, heat exchange can be arranged between the in and out flows of oil.

Continuous plants are similar in operation but the oil flows continuously through the main bleaching vessel. One of the main functions of the bleaching process is the reduction of the trace metal content of the oil, in particular of iron and copper. Wide use is made of stainless steel in the construction of continuous and semicontinuous plants.

Spent earth is removed from the oil by the use of various types of filter. Examples are:

1. plate and frame presses with open or closed discharge;
2. chamber presses, sometimes fitted with diaphragms for squeezing the oil from the spent filter cake;
3. vertical or horizontal tank filters with vertical leaves of stainless steel mesh;
4. vertical tank filters with horizontal leaves attached to a central shaft which can be rotated for cake discharge.

All of these filters are, preferably, backed up by a polishing or 'guard' filter to prevent the accumulation of spent earth in the bleached oil storage tanks, since oxidation products on such spent earth promote further oxidation of the stored oil.

4.3.4 Physical refining pretreatment

The physical refining process consists of two major stages, pretreatment and distillation. In the second of these (Table 4.7), the bulk of the free fatty acids is removed, the carotenoids are substantially decomposed, and the volatile decomposition products are distilled from the oil. There is also a reduction of the content of oxidation products and monoglycerides. The object of the pretreatment process is therefore:

1. to reduce the phosphorus level to a maximum of 4 mg/kg of oil;
2. to reduce trace iron and copper contents to at most 0.2 and 0.06 mg/kg respectively;
3. to reduce the content of pigment decomposition products which are not heat decomposable and volatile;
4. significantly to reduce the content of oxidation products.

Many chemicals have been tried for pretreatment of palm oil, but only one process has found widespread use. This involves the use of phosphoric acid, in some cases together with citric acid and activated bleaching earth. The phosphoric acid decomposes the calcium and magnesium compounds that form the non-hydratable phosphatides, and coagulates the phosphatides rendering them insoluble in the oil. The iron and copper complexes are attacked in the same reaction and are removed together with the phosphatides, pigments and a part of the oxidation products by adsorption on the earth. Some refiners add calcium carbonate to the oil after the earth in order to form insoluble calcium phosphate by reaction with free phosphoric acid in the oil. The phosphate and earth are filtered from the oil.

The process can be carried out in batch, semicontinuous or continuous equipment. The semicontinuous plant is preferable because the plant can be designed to give a minimum finite time of contact between the chemicals and the oil. A modification of the bleaching plant (Figure 4.4) is a good example of the pretreatment equipment, which consists of a unit fitted upstream of the main bleaching vessel. In this unit the oil is deaerated under vacuum, treated with 0.05–0.10 per cent of 80% phosphoric acid, heated to 90–110 °C and given a reaction/residence time of 15 min before passing to the bleaching vessel. There,

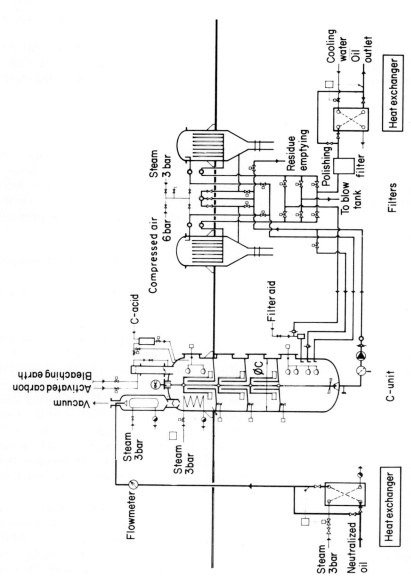

Figure 4.4 Semicontinuous bleaching plant (Reproduced by permission of AB Pellerin-Zenith).

1.0–1.5 per cent of activated bleaching earth is added to the oil followed, if desired, by calcium carbonate stoichiometrically equivalent to the original phosphoric acid addition. The minimum bleaching time is 15–20 min, after which the oil is filtered, polished and cooled.

In a modification of this process oil at 50 °C is treated with 0.05–0.10 per cent of 80% phosphoric acid, heated to 90 °C, when 5–10 per cent of hot, softened water is added. The aqueous phase, which contains the bulk of the phosphatides and excess phosphoric acid, is separated centrifugally and the oil is dried under vacuum before being bleached and filtered as before.

4.3.5 Deodorization and distillation

(a) Introduction Deodorization is the final major stage in the refining of edible oils. Its purpose is to produce a product of bland flavour and odour and of good shelf-life stability, so the process must remove, as far as possible, free fatty acids and the aldehydes and ketones (formed by oxidation of the unsaturated fatty acids) principally responsible for the unacceptable taste and smell of the oil after bleaching. Also removed are pigments (mainly carotenoids by thermal breakdown), tocopherols and tocotrienols and other unsaponifiable compounds such as hydrocarbons and alcohols. The process used is steam distillation under vacuum.

The distillation stage of physical refining is a special case of deodorization. The objectives are the same but the quantity of free fatty acids to be removed in the case of palm oil is greater, generally between 3 and 6 per cent. The ability of the deodorizing step to produce an oil of good quality is limited by the quality of the crude oil, the processing and storage treatment of the oil prior to deodorizing and the deodorizing plant itself.

The process does not reduce trace metals or phosphatide levels. If the oil has been neutralized with caustic soda, soap removal should be complete at the bleached and filtered stage; otherwise high losses will be encountered due to foaming in the deodorizer. The deodorizer will only reduce the Totox value (Totox value, $TV = 2 \times$ peroxide value + anisidine value) by about 50 per cent and therefore contact with air in prior-refining equipment must be minimized, and during storage of the bleached or pretreated oil the temperature should be held at about 10 °C above its melting point.

(b) Process description Plant designs available for deodorization are batch, continuous, semicontinuous and continuous–semicontinuous. Batch plants can only be used for palm oil if they are specially equipped for high-temperature operation or the oil has been decolourized prior to deodorization. They are not used for physical refining. In continuous operation the oil flows in an uninterrupted manner through the successive stages of the process. In the semicontinuous method, small batches of oil are treated to the individual process steps in purpose-designed parts of the plant. The continuous operation is simple and lends itself to heat-recovery practices while the semicontinuous process is less energy efficient but gives oil a defined residence time in the distillation stages. The continuous–semicontinuous plant has the advantages of both of the two previous designs. The following process description describes such a design (Figure 4.5). Plant capacity lies generally between 50 and 360 t/d.

(i) Deaeration Because of the necessity for protecting the oil against oxidation at all stages and the increase in rate of oxidation with increasing temperature, it is essential to remove air from the oil prior to heating. This is done either by pumping the pretreated oil to the first heating stage through a deaerating vessel connected to plant vacuum (as shown in Figure 4.5) or by cascading the oil as a thin film into the deodorizer which is under vacuum.

(ii) Heating Following deaeration, oil is heated to the required process temperature. In recent years, heat exchange procedures have been introduced which can, depending on the type of plant, save 50–80 per cent of heating costs.

The oil temperature after heat exchange will vary between 160 and 220 °C depending on the plant and process conditions used. Final heating to the required process temperature is effected by high-pressure steam or other heat exchange medium such as thermal oil or a eutectic mixture of diphenyl and diphenyl oxide. When carried out in an internal section or tray of the deodorizer (Figure 4.5), sparging steam is used to improve efficiency of heat exchange, prevent burning of the product and initiate the deodorization process. The oil overflows from the first section, tray 1, into tray 2 where a batch is accumulated with steam injection and is dropped by programmed operation of the tray drop valve into tray 3, which is the main deodorizing/distilling section.

(iii) Deodorization/distillation The deodorization process is concerned with system pressure, temperature, steam to oil ratio and time. From the theory of the process[24], halving the system pressure halves the requirement for stripping steam. However, because of energy costs involved in achieving high vacuum, most commercial plants operate at 2.5–5.0 mmHg abs. To reduce the chance of refluxing the vacuum should be the same in the various stages of the process.

The deodorization and deacidification temperature depends on the vapour pressure of the compounds to be removed. Although carotene is substantially destroyed and removed at 240 °C, in practice higher temperatures of up to 270 °C are required to break down compounds formed by oxidation during storage. Temperatures are kept as low as possible to reduce losses and the possibility of isomerization and other thermochemical reactions[25–27]. It is in the interests of the refiner to minimize the loss from the oil of natural antioxidants, tocopherols and tocotrienols. There is a breakeven point at which

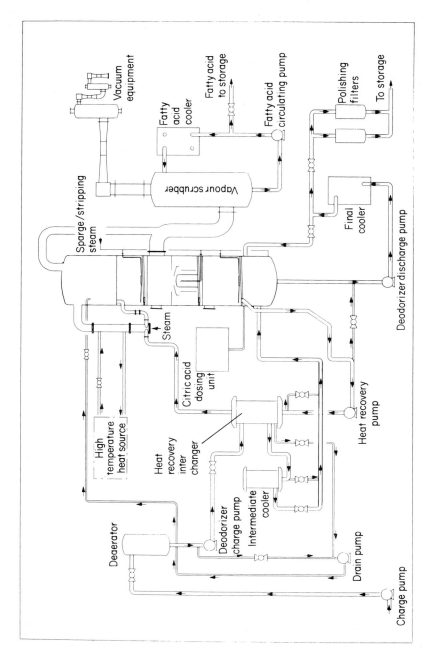

Figure 4.5 Continous/semicontinuous Econonoflow deodorizing plant. (Reproduced by permission of Simon Rosedowns Limited).

the advantage to shelf life of removal of aldehydes and ketones is cancelled out
by the loss of these antioxidants. In commercial practice the temperature of
operation lies between 240 and 270 °C, depending on the quality of the oil and
the plant design.

The volume of steam required for deodorization is directly proportional to the
system pressure and inversely proportional to the vapour pressure of the FFA.
Thus, a reduction in the former and an increase in the latter, with increasing
temperature, results in a reduction of time at temperature for a set steaming
rate. This is correct for the simple removal of fatty acids. However, oils vary in
their content of pigments and oxidation products, and practical experience has
shown that, while these products can be removed in the time required to reduce
FFA to the desired level from a good quality feed oil, this is not so with oxidized
oils. For such oils, an extended time at temperature is required to allow thermal
reactions to take place in which some oxidation products are further decomposed and the derivatives removed. If such oxidation products are not removed,
the product will have a poorer taste and reduced oxidative stability. The
limitations of this aspect of the deodorization process can be seen by noting that
the anisidine value (a measure of secondary oxidation products in the oil) is not
reduced to zero. Commercial plants are currently designed for holding time at
temperature of 30–120 min, but all are capable of extension. The live steam
requirement is normally between 1 and 4 per cent of the oil flow.

(iv) Cooling In modern plant the first stage of cooling is achieved by heat
exchange against ingoing feed oil. Cooling is continued from 120–150 °C by
heat exchange against water, either internally under vacuum with stripping
steam or externally against feed oil and/or water. The temperature of the
cooled oil is usually 60–65 °C. For stability reasons further cooling to 35 °C for
palm olein or 10 °C above the melting point for other palm products is
recommended. Citric acid, for chelation of trace metals, and antioxidants, if
required, are added during the cooling process at about 120 °C, or to the cooled
oil.

In the plant shown in Figure 4.5 one or two distillation trays are installed in
each of which the oil has a defined residence time. When this time has elapsed
the oil is automatically dropped into the heat-recovery tray from where it is
continuously pumped through the interchanger and back into the main vessel.
Volatile oxidation products formed during the cooling step are removed and
0.01 per cent of citric acid in aqueous solution is added to the oil before it is
pumped to storage via final cooling. The polishing filter is needed to remove
traces of dirt in the oil.

(v) Environmental factors[28] Entrained neutral oil is carried to the
vacuum equipment and thence via the condenser water to the drain. To reduce
this contamination significantly a vapour scrubber (fat eliminator) is placed
between the deodorizer and the vacuum equipment, using co- or countercurrent

spraying of fatty acids and demisting devices to condense and coalesce fatty material in the vapour. The level of fat in the condenser water is reduced from about 300 ppm to approximately 25 ppm by this means.

To save water, the condenser water is normally recycled via cooling towers. The cooling tower water gives off the objectionable smelling compounds taken from the oil. To overcome this air pollution problem some refiners have installed a double-loop system in which pure water is circulated through one of two plate heat exchangers used alternately where the condenser water is cooled before being returned to the vacuum equipment. The pure water is cooled in the cooling tower which thus remains clear of fat. The heat exchanger must be cleaned periodically with hot water or hot caustic soda solution.

4.3.6 Refining after transportation

A very large proportion of the world's palm oil production is transported long distances after it has been refined and deodorized. Sometimes the oil becomes contaminated, principally with iron, during transportation, or for other reasons arrives at its destination unfit to be used or placed in the market. Depending on the condition of the oil, degrees of further refining treatment are then required. These are as follows:
1. Filtration for those oils which are dirty but otherwise satisfactory.
2. Filtration and deodorization for oils of poor flavour or high oxidation values or high free fatty acid content.
3. Treatment with citric acid and activated bleaching earth followed by filtration and deodorization for poor quality oils known to have high contents of iron. Citric acid as a chelating agent assists in the removal of iron.
4. In very bad cases neutralization, bleaching and deodorization are required.

4.3.7 Effect of refining on oil composition

In Table 4.8 specifications are given for the quality of fully refined palm oil at the time of initial production. The reductions effected by refining can be seen by comparison of these figures with those given in Table 4.6. Although not stated, the oil should also be odourless and of bland flavour. The PORAM specification is a minimum requirement with other parameters being agreed between the buyer and seller. Jacobsberg[14] under 'consumers' appreciations' gives, for the AOM test, $\geqslant 40$, ± 30 and <20 h for good, medium and poor quality oils respectively. The Totox values for the same oils in the same order are given as 8 maximum, ± 20 and >30.

During bleaching using the quantities of activated earths mentioned in this chapter, the PV is reduced to nil and the TV to about half of its value in the crude oil. During deodorization or the distillation step of physical refining there should be a further 50 per cent reduction in the TV, making 75 per cent in all. The TV of a freshly deodorized oil from good average crude oil should therefore be less than 8.

Table 4.8 Specifications for fully refined palm oil

	PORIM recommendations	PORAM[a] as loaded	Consumer's specification of good quality[27]
Free fatty acid, palmitic (%)	0.05 max.	0.1 max.	0.05 max.
Iodine value (Wijs)	—	50–55	—
Slip melting point (°C) (AOCS Cc 3-25)	—	33–39	—
Colour, 5.25 in Lovibond	—	3 or 6 Red	3 Red max.
Phosphorus (mg/kg)	4 max.	—	4 max.
Iron (mg/kg)	0.12 max.	—	0.1 max.
Copper (mg/kg)	0.05 max.	—	0.05 max.
AOM stability (h)[b]	50 min	—	—
PV (meq/kg)	Nil	—	—
Anisidine value	8 max.	—	—
Totox value	8 max.	—	10 max.
Dimer (%)	—	—	1 max.
Moisture and impurities (%)	—	0.1 max.	—

[a]Palm Oil Refiners' Association of Malaysia.
[b]Active oxygen method for fat stability AOCS Cd 12-57.

Dimerization of oxidized or unsaturated triglycerides will occur if conditions of heating and time spent at the high temperature are excessive. The Unilever opinion[25] is that the dimer content of the refined oil should be less than 1 per cent. To achieve this level it is recommended that palm oil should be deodorized at a maximum of 240 °C for a maximum residence time of 2 h. It is suggested[27] that 270 °C for a maximum of 30 min might be acceptable.

Examples of tocopherol and tocotrienol losses of between 15 and 57 per cent across the whole refining process were given by Maclellan[8]. The considerable variation in the figures results from the process conditions and plant employed and the degree of care taken during intermediate storage.

It must be pointed out that the oxidation which takes place during transportation and any consequent deodorization of the oil on arrival at its destination will further decrease the tocopherol/tocotrienol content, and the latter will also tend to increase the dimer level.

4.4 Fractionation

4.4.1 General considerations

Oils and fats are mixtures of triglycerides which, because of their different fatty acid compositions, have melting points spanning the range from below −30 °C to above +70 °C, each oil having its own melting range. The melting range, which is measured by determining the solid fat content curve (Table 4.4) limits the use of a particular oil or fat. Fractionation is a thermomechanical process by which the raw material is separated into two or more portions which widens its use.

Thermomechanical separation processes include distillation and crystallization. Distillation is commercially unsuitable for the fractionation of triglyceride mixtures because of their low vapour pressure and because of their relatively low stability at high temperatures. Separation can, however, be effected by crystallization.

The triglyceride composition of palm oil (Table 4.3) includes substantial quantities of both low and high melting point triglycerides. The oil therefore lends itself to crystallization by controlled cooling followed by separation to yield a liquid (olein) and a solid (stearin) phase.

The aim of single fractionation of palm oil is to produce an olein which can be used as a partial replacement for liquid vegetable oils such as soyabean oil. The stearin from this fractionation is used in margarine, shortening and frying fat production. If a double fractionation is carried out in which, for example, the olein from a first stage is recrystallized at a lower temperature, a second stearin and olein are obtained. The second stearin is known as palm mid-fraction (PMF). The temperature of the two stages of the process can be so chosen that the PMF is enriched in the symmetrical triglycerides POP and POS, which are important in the production of cocoa butter equivalents. The importance of fractionation in palm oil processing can be seen in that over 2 million t of the oil were fractionated in 1983 out of a total world production of 6.4 million t[29].

Edible oils and fats are polymorphic in their crystalline behaviour. When considering the crystallization of palm oil three forms are of importance, α, β' and β, which are of increasing stability in that order. The rate of crystallization of the α form is greater than that of the β', which in turn is greater than that of the β polymorph. If supercooling is carried out too rapidly, crystallization of the α form occurs resulting in a mass of very small crystals. To obtain good separation of palm oil fractions, crystallization is required in the β' form. This is because the β' crystals agglomerate into large aggregates that are firm and of uniform spherical size and give good filtration.

Crystallization is affected by the non-triglyceride components of edible oils. Free fatty acids form eutectic mixtures with the triglycerides causing a softening of the consistency of the fat[30]. This feature is indicated by an increase in the solids content of palm olein after physical refining when the original palm oil is fractionated in the crude state.

As stated earlier, the monoglyceride content of palm oil is normally less than 1 per cent. At this level monoglycerides have no noticeable effect on crystallization or separation but Koslowski[6] reported inhibition of crystallization when working with an oil of 9 per cent free fatty acid and 5.5 per cent monoglyceride. Diglycerides, on the other hand, are or can be present in sufficient quantities in commercial oils to affect crystallization. Their effect is caused by the formation of eutectic mixtures and by slowing down the transformation of the α to the β' form. This latter effect causes difficulties with filtration possibly due to differences in the size and shape of the crystals obtained[30]. Other impurities such as phosphatides, sugars and soap inhibit the process either by increasing the viscosity of the olein phase or by making the crystal agglomerates slimy.

Table 4.9 Concentration of minor components after fractionation

Olein	Stearin
Free fatty acids	Iron
Carotene	Phosphorus
Tocopherols	Partial glycerides
Partial glycerides	(hexane)
(acetone)	

In the fractionation process the minor components of the original oil become concentrated in the fractions (Table 4.9). This concentration has a considerable effect on the oxidative stability of the fractions. Relative to the starting oil, the olein is enriched in tocopherols and depleted in iron: the reverse occurs with stearin, which therefore becomes appreciably more susceptible to oxidation despite its lower content of unsaturated acyl groups. The situation is worsened since, because of its high melting point, the stearin must be stored and handled at a considerably higher temperature than the olein.

The importance of these opposing results on partial glyceride content after solvent fractionation in acetone and hexane lies in the possibility that a palm mid-fraction from hexane could contain significant quantities of partial glycerides to the detriment of its performance in a cocoa butter equivalent. On the other hand, high levels of partial glycerides in palm olein degrade the product for use as a frying oil. Willems and Padley[27] discuss the importance of oil quality and processing parameters for the production of palm mid-fraction.

Quality testing of the products from fractionation is conducted according to the purposes for which they are intended. In many cases iodine values and melting points are sufficient in addition to the common general quality tests. Palm olein is assessed by its cloud point and sometimes by a 'cold stability' test in which the length of time for which the oil remains clear is measured at a stated temperature, usually between 16 and 22 °C. Other tests of importance, primarily for stearin, are the differential scanning calorimetry curve and the solid fat content by n.m.r. or the solid fat index by dilatometry. Deffense and Tirtiaux[31] have shown the value of high-performance liquid chromatography for monitoring the raw materials and products of palm oil fractionation.

4.4.2 Process description

There are three processes in commercial use for the fractionation of palm oil. They are referred to as the dry, the detergent and the solvent processes. In the dry system the oil is crystallized and separation is effected by filtration. In the detergent process the oil is again crystallized on its own but separation is effected employing an aqueous detergent solution and centrifugation. The solvent process employs crystallization and filtration in miscella. Of the three by far the largest tonnage is processed by the first two methods. Maclellan[8] stated that in Malaysia the installed daily capacity for the dry, detergent and solvent processes was 4950, 4550 and 200 t respectively. Because of its operational cost the solvent process is now restricted to the production of a POP-rich fraction for incorporation in cocoa butter substitutes/equivalents. Process descriptions will therefore be restricted to the dry and the detergent systems.

(a) Dry fractionation[31] The process can be carried out using as raw material either crude, pretreated, neutralized and bleached or fully refined palm oil (Figure 4.6). The oil is first heated to 70–75 °C so as to melt all crystal nuclei and then passes to a crystallizer. The number and size of the crystallizers depends on the throughput of the plant and on the mode of operation. In some installations the full crystallization is carried out in one tank; in others the cooling is conducted in stages with one crystallizer per stage and the oil is pumped from one crystallizer to the next through the various stages. The plant in Figure 4.6 is of the first type. Continuity of supply to the filter is obtained by using the number of crystallizers appropriate for the throughput.

The crystallizer is a tall cylindrical vessel equipped for jacket cooling, with an agitator designed to scrape solidifying fat from the tank's inner surface. The agitator speed depends on design but is usually between 5 and 15 r/min. After initially cooling to 45 °C in about 45 minutes, further cooling to the filtration temperature of 18–20 °C takes 4–8 h, depending on the plant design. The crystalline slurry is pumped to a filter for separation. Filters used for the purpose are of the plate and frame, rotary vacuum and vacuum belt designs. With these filters the yield of olein is in the 60–70 per cent region, but recent experience with membrane chamber filters, in which pressure on the membrane is used to squeeze olein from the stearin filter cake, has shown that 75–80 per cent yield is attainable.

(b) Detergent fractionation[32] The crystallization step of this process (Figure 4.7) is similar to that described for dry fractionation but the crystal form is less critical because of the means of separation used. A crystallization time of less than four hours is adequate. A continuous supply of crystalline slurry is pumped to the plant. The slurry is mixed with a detergent solution, usually sodium lauryl sulphate, and an electrolyte (magnesium sulphate). The solutions displace the oil from the surface of the stearin crystals so that, on centrifuging, the olein is discharged as the oil phase and the stearin forms part of the aqueous phase. The aqueous phase is heated to 95–110 °C to melt the stearin and break the emulsion, and it is then cooled to 90 °C and centrifuged to recover the stearin and the detergent/electrolyte solution which is recycled.

Regulations in force in certain countries limit the allowable amounts of sodium lauryl sulphate in fractionated oils. The limits are achievable by washing or refining of the fractions, but they restrict the range of raw materials to which the process can be applied by comparison with the dry process. The olein yield at 77–83 per cent is high.

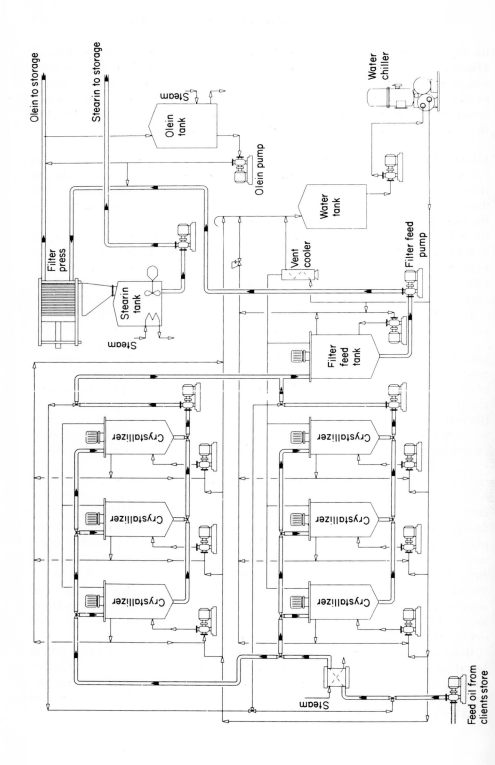

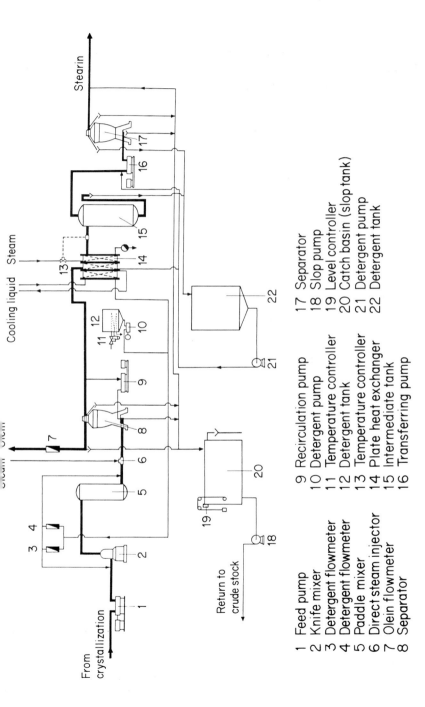

From crystallization

Steam

Cooling liquid   Steam

Stearin

Return to crude stock

1 Feed pump
2 Knife mixer
3 Detergent flowmeter
4 Detergent flowmeter
5 Paddle mixer
6 Direct steam injector
7 Olein flowmeter
8 Separator

9 Recirculation pump
10 Detergent pump
11 Temperature controller
12 Detergent tank
13 Temperature controller
14 Plate heat exchanger
15 Intermediate tank
16 Transferring pump

17 Separator
18 Slop pump
19 Level controller
20 Catch basin (slop tank)
21 Detergent pump
22 Detergent tank

Figure 4.7 Detergent fractionation plant. (Reproduced by permission of Alfa Laval AB).

4.4.3 Products

Table 4.10 gives values for some of the characteristics of oleins and stearins derived from the single fractionation of palm oil. The Malaysian Standard specifications are minimum requirements. Buyers must agree with the seller on such parameters as cloud point, melting point and solid fat content.

Table 4.10 Palm oil single fractionation: analyses of refined oleins and stearins

	Olein			Stearin	
	MS 816:1983[33]	Range [33, 34]	Mean [33, 34]	MS 815:1983[35]	Range [34, 35]
FFA, palmitic (%)	0.10 max.	—	—	0.20 max.	—
Moisture and impurities (%)	0.10 max.	—	—	0.15 max.	—
Colour, Lovibond 5.25 in	6 Red max.	—	—	6 Red max.	—
Iodine value (Wijs)	56 min.	56.1–60.6	58.0	46 max.	21.6–49.4
Melting point (°C) AOCS Cc 3-25	24 max.	19.4–23.5	21.6	48 min.	44.5–56.2
Cloud point (°C) Refined, AOCS Cc 6-25	—	6.0–11.5	8.8	—	—
Solid fat content (%)					
5°C	—	43.6–61.0	51.1	—	63.3–91.6
10	—	28.1–51.8	37.0	—	54.2–91.1
15	—	13.3–24.9	19.2	—	41.9–90.9
20	—	2.9–8.6	5.9	—	31.3–87.4
25	—	—	—	—	20.2–81.9
30	—	—	—	—	16.4–73.5
35	—	—	—	—	12.5–65.0
40	—	—	—	—	7.4–56.6
45	—	—	—	—	2.7–48.6
50	—	—	—	—	0–39.7
55	—	—	—	—	0–19.3

Olein values fall into a relatively narrow range, but there is a wide divergence in stearin values. Stearins from the detergent process tend to be of higher melting consistency than those from the dry process because the separation is cleaner in the former case. In dry fractionation, a certain amount of olein is occluded in the stearin filter cake. The lower melting stearins are, however, generally more popular commercially because they are easier to use in blends for margarines and similar fat products. Deffense[29] compares the oleins and stearins obtainable by the different methods, using the equipment of individual plant suppliers, and also includes information on solid fat content and fatty acid and triglyceride composition.

The same paper gives full analytical values for three 'super oleins' and three palm mid-fractions from double-stage fractionations. The 'super oleins' have iodine values from 60.7 to 62.7 and cloud points from 4.8 to 5.7°C.

Palm mid-fractions produced by the dry and the detergent processes are not suitable for incorporation into cocoa butter substitutes as they stand but must be 'refined' by further solvent fractionation to eliminate di- and triunsaturated triglycerides.

The analytical criteria of a mid-fraction as currently defined by PORIM in conjunction with the Malaysian industry are:

Parameter	Value
Ratio $\dfrac{C_{50}}{C_{48} + C_{54}}$	4 min.
C_{52} (%)	43 max.
Iodine value (Wijs)	32–55
Slip melting point (°C)	23–40

4.5 References

1. Codex Standard 125-1981, Codex International Standard for Edible Palm Oil, FAO/WHO.
2. M. Loncin, B. Jacobsberg and G. Evrard, 'Palm oil, a major tropical product', Tropical Product Sales, Unilever House, Brussels (1970).
3. B.K. Tan and C.H. Oh Flingoh, 'Malaysian palm oil chemical and physical characteristics', PORIM Technology 3 (1981).
4. J.A. Cornelius, J. Am. Oil Chem. Soc., 54, 943A-8A (1977).
5. G. Jurriens, B. De Vries and L. Schouten, J. Lipid Res., 5, 36 (1964).
6. L. Koslowski, Grasas y Aceites, 26, 95 (1975).
7. B. Jacobsberg and C.H. Oh, J. Am. Oil Chem. Soc., 53, 609–17 (1976).
8. M. Maclellan, J. Am. Oil Chem. Soc., 60, 368–73 (1983).
9. J.B. Rossell, B. King and M.J. Downes, J. Am. Oil Chem. Soc., 60, 333–9 (1983).
10. S.H. Goh, H.T. Khor and P.T. Gee, J. Am. Oil Chem. Soc., 59, 296–9 (1982).
11. S.H. Goh, Y.M. Choo and S.H. Ong, J. Am. Oil Chem. Soc., 62, 237–40 (1985).
12. S.H. Goh, S.L. Tong and P.T. Gee, J. Am. Oil Chem. Soc., 61, 1597–600 (1984).
13. S.H. Goh, S.L. Tong and P.T. Gee, J. Am Oil Chem. Soc., 61, 1601–4 (1984).
14. B. Jacobsberg, 'Quality of palm oil', PORIM Occasional Paper 10 (1983).
15. W.L. Leong and K.G. Berger, 'Storage, handling and transportation of palm oil products', PORIM Technology 7 (1982).
16. F.V.K. Young, Proceedings of the 2nd American Soybean Association Symposium on Soybean Processing, Antwerp, Belgium (1981).
17. G. Haraldsson, J. Am Oil Chem. Soc., 60, 251–6 (1983).
18. L. H. Wiedermann, J. Am Oil Chem. Soc., 58, 159–66 (1981).
19. J.C. Segers, J. Am. Oil Chem. Soc., 60, 262–4 (1983).
20. U.I. Brimberg, Fette, Seifen Anstrichmittel., 83, 184–90 (1981).
21. U.I. Brimberg, J. Am. Oil Chem. Soc., 59, 74–8 (1982).
22. W. Zschau, Fette. Seifen Anstrichmittel., 84, 493–8 (1982).
23. T.K. Mag, United States Patent 4230 630 (1980).
24. A.M. Gavin, J. Am. Oil Chem. Soc., 55, 783–91 (1978).
25. S.R. Eder, Fette. Seifen Anstrichmittel., 84, 136–41 (1982).
26. J.B. Rossell, S.P. Kochhar and I.M. Jawad, Proceedings of the 2nd American Soybean Association Symposium on Soybean Processing, Antwerp, Belgium (1981).
27. M.G.A. Willems and F.B. Padley, J. Am. Oil Chem. Soc., 62, 454–9 (1985).
28. W.J. Gilbert and D.C. Tandy, J. Am. Oil Chem. Soc., 56, 654A–8A (1979).
29. E. Deffense, J. Am. Oil Chem. Soc., 62, 376–85 (1985).
30. K.G. Berger, Oil Palm News, 22, 10–18 (1977).
31. E. Deffense and A. Tirtiaux, Paper presented at American Oil Chemists' Society conference,

The Hague, Netherlands (October 1982). Address: S.A. Fractionnement Tirtiaux, 601 Chaussée de Charleroi, B-6220 Fleurus, Belgium.
32. G. Haraldsson, Paper presented at seminar, Kuala Lumpur, Malaysia (29 April 1978). Address: Alfa Laval AB, PO Box 14700, Tumba, Sweden.
33. Malaysian Standard, 'Specification for palm olein', MS 816:1983.
34. B.K. Tan and C.H. Oh Flingoh, 'Oleins and stearins from Malaysian palm oil, chemical and physical characteristics', PORIM Technology 4 (1981).
35. Malaysian Standard, 'Specification for palm stearin', MS 815:1983.

5   End uses of palm oil

Human food

S.A. Kheiri

5.1	**Introduction**	71
5.2	**Processed palm oil products**	72
5.3	**Food uses**	74
5.3.1	Oils and fats for cooking	77
5.3.2	Fats for bakery products	78
5.3.3	Oils/fats for table margarine	80
5.3.4	Confectionery fats	82
5.3.5	Miscellaneous	83
5.4	**References**	83

5.1 Introduction

Before the Second World War palm oil was employed only as a technical oil. Thereafter, especially since 1952, much research and development in the production and processing of palm oil have been carried out in the producing countries, notably in Malaysia, the world's largest producer and exporter of the oil, to improve the quality of palm oil and its products. As a result of these concerted efforts which resulted in tremendous improvement in the quality of palm oil, more than 90 per cent of the world consumption of palm oil is now used in human food. This is equivalent to about 13 per cent of the world consumption of oils and fats used in food products for human consumption.

The United Kingdom was the first country to import palm oil for human consumption outside producing countries and now uses more than 100 000 tonnes of palm oil in food products each year. Because of the import duty on palm oil and the availability of soyabean and rapeseed oils from local seed crushers, the use of palm oil in Britain has declined over the years, as shown in Figure 5.1. In 1980 about 90 per cent of the palm oil imported into the United Kingdom was used for human consumption. Only 10 per cent was used for technical purposes.

It should therefore be recognized that palm oil, due to its consistent quality, technical interchangeability and price competitiveness with other edible oils and

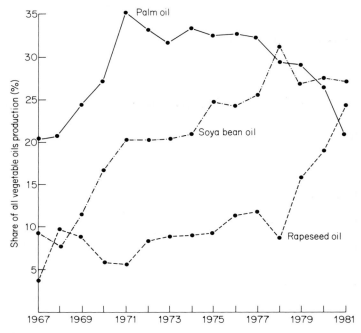

Figure 5.1 Production shares of major oils in the United Kingdom. Vegetable edible oil market 1967–81[1].

fats, is now essentially an edible oil and as such is used, both in the producing and importing countries, mainly in foods for human consumption. Its use in products other than food, as with other edible oils and fats, is price sensitive and varies depending on the price of tallow and other technical oils and fats.

World production of oils and fats used in human foods is given in Table 5.1. Palm oil now commands the second position in terms of percentage share of total world production of oils and fats. In view of the increased acreage under palm oil in some South-East Asia, African and South American countries this percentage share is likely to increase in the future. About 60 per cent of the total production of palm oil is exported and used mainly in human food by non-producing countries. The critical role of palm oil as one of the major supplier of oils and fats for human food cannot, therefore, be overemphasized. The amounts of palm oil and its fractions imported mainly for human consumption by selected countries from Malaysia is shown in Table 5.2.

5.2 Processed palm oil products

To increase the domestic added value and to find new markets for their oil, some palm oil producing countries, Malaysia in particular, have established plants to process the oil before exporting to the consuming countries. As a consequence,

Table 5.1 World production of oils and fats

Oils/fats	Share held during the year (%)		
	1970	1976	1982
Annual seed crops			
Soyabean oil	16.6	21.8	22.8
Rapeseed oil	5.2	6.4	7.3
Sunflower seed oil	10.5	7.9	9.2
Groundnut oil	9.0	7.7	6.1
Cotton seed oil	6.6	5.9	6.2
Total	47.9	49.7	51.6
Perennial crops			
Palm oil	4.7	6.5	10.8
Palm-kernel oil	1.2	1.2	1.5
Coconut oil	5.8	7.3	5.5
Olive oil	3.5	3.9	2.1
Total	15.2	18.9	19.9
Animals			
Tallow	12.2	11.7	10.7
Lard	11.3	7.3	6.7
Fish oil	2.8	2.1	2.4
Butter	10.6	10.3	8.7
Total	36.9	31.4	28.5

Source: Oil World, 1983.

Table 5.2 Processed palm oil products imported by selected countries from Malaysia, mainly for human consumption

Country	Amount (t)							
	1982				1983			
	Palm Oil	Olein	Stearin	Total	Palm Oil	Olein	Stearin	Total
Australia	299	37 132	15 222	52 653	1 237	63 616	14 527	79 380
West Germany	17 732	6 366	17 014	41 112	12 024	1 495	9 847	23 366
India	154 791	245 631	90	400 512	282 710	302 488	8	585 166
Japan	50 486	54 931	20 703	126 120	71 524	51 973	16 521	140 018
Netherlands	91 191	15 814	50 933	157 938	74 068	7 650	35 831	117 549
Pakistan	258 130	4 017	—	262 147	326 482	12 036	6 033	344 551
United Kingdom	16 858	35 530	7 561	59 949	8 218	18 657	3 507	30 382
United States	51 366	16 720	28 352	96 438	101 824	23 289	19 703	144 816
USSR	164 919	—	86 651	251 570	176 158	—	80 280	256 438

Source: Department of Statistics, Malaysia.

about 50 per cent of the world export of palm oil is now traded as partially and fully processed palm oil and its fractions, olein and stearin.

Palm oil, olein and stearin are respectively semisolid, liquid and solid at 20 °C. Solid fat contents of typical palm oil, olein and stearin exported from Malaysia are shown in Figure 5.2. Both palm oil and olein are, within reasonable limits, quite consistent in their composition and properties. Palm stearins, on the other hand, are produced and exported[3] with iodine values ranging from 30 to 48. The availability of these palm oil processed products has enabled users to choose a product most suited to their application. It has also made it possible, especially in countries with a shortage of processing capacity, to use these products in human food without further processing.

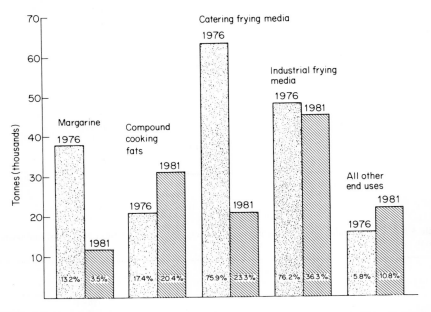

Figure 5.2 Tonnages of palm oil used in main food uses and percentage share of edible oils/fats in the United Kingdom[2].

5.3 Food uses

Almost all oils and fats, to one degree or another, are interchangeable in most applications, including human foods. Improved techniques of fractionation, hydrogenation and interesterification have widened the performance of most oils and fats to a point where they are virtually interchangeable, partially or wholly, in any application. The inherent technical properties, to a certain extent, are now irrelevant in most food products. Oils and fats or their blends are chosen according to the prices and costs involved in processing and making them suitable for specific applications.

Palm oil has a range of distinctive properties which enable it to meet some of the more demanding quality requirements in edible products. On the other hand, some of its characteristics restrict its use to low levels in certain applications.

Rourke[4] summarized palm oil characteristics which are important in determining its incorporation in food products as follows:

Advantages

1. Natural colouring materials can provide a colourant for margarine and yellow fat.
2. A high solid glyceride content gives consistency without hydrogenation.
3. Low linoleic and linoleic acid contents give good heat stability.
4. A low level of triglycerides containing short chain fatty acids which are susceptable to hydrolysis, thereby minimizing development of off-flavour from microbial action.
5. The level of high melting point triglycerides combined with relatively low solid content at 10 °C helps in the formation of products with a wide plastic range, suitable for hot climates and some industrial applications.

Disadvantages

1. A high carotenoid level, especially when colouring agents are modified during oxidation, may make it difficult to produce a low coloured oil, or high costs may be incurred in reaching the acceptable level of colour.
2. Linoleic acid in the region of 10 per cent, although making a useful contribution to the essential fatty acid content of margarines, represents too low a level to allow appreciable quantities of palm oil to be incorporated in margarines specifying a high polyunsaturated fatty acid level.
3. The wide plastic range gives relatively poor melting in the mouth.
4. The free fatty acid content increases rapidly in overripe or damaged fruit or in crude or processed oil that is incorrectly prepared for storage or is held under satisfactory conditions. This can cause high refining costs.
5. The high free fatty acid content facilitates the pick-up of prooxidant contaminants that reduce the inherent stability of the oil during storage and during heavy duty applications.
6. Slow crystallization properties can promote structural hardness in finished products and also aggravate a tendency for recrystallization to occur with a consequent impairment of texture.

Disadvantages 1, 4 and 5 do not apply to oil that has been refined and deodorized in the producing country before export. Other disadvantages can be overcome by modifying palm oil, employing such processes as fractionation, hydrogenation and interesterification.

Important characteristics of palm oil and some of its modified products used in human food are shown in Table 5.3. These products are now widely used not only in countries producing palm oil but also in those importing the oil for human consumption. As an example, the amount of palm oil used in the United Kingdom in 1976 and 1981 in various food products are shown in Figure 5.2.

Table 5.3 Characteristics of palm oil and its products used in foods

Parameters	PO	HPO	Int.PO	POo (x1)	POo (x2)	HPOo	Int.POo	POs (soft)	POs (medium)	POs (hard)	PMF (A)	PMF (B)
Iodine value	54.3	44.6	—	58.4	61.4	43.8	61.0	47.6	43.7	32.9	40.1	42.5
Slip melting point (°C)	36.9	41.5	46.3	22.2	—	42.5	40.6	44.2	48.4	53.0	—	—
Trans value	—	11.3	—	—	—	15.8	—	—	—	—	—	—
Solid fat content (%)												
5°C	74.8	82.1	62.8	54.0	—	85.9	55.0	—	70.5	—	—	—
10°C	52.0	76.2	52.3	39.7	16.7	81.8	45.1	63.9	65.9	82.7	82.4	76.1
15°C	38.7	66.5	41.8	21.6	5.0	76.9	33.1	52.3	58.2	78.0	71.8	66.2
20°C	25.2	53.6	29.5	5.9	3.1	67.3	23.8	39.9	47.6	72.2	61.3	52.1
25°C	16.2	39.3	25.9	2.7	—	52.1	15.8	29.7	36.1	64.8	31.6	31.9
30°C	9.7	26.9	22.8	0	—	36.2	11.0	21.4	26.5	55.9	8.6	18.3
35°C	7.4	16.5	15.0	0	—	23.9	8.3	16.5	21.8	49.2	4.0	13.4
37.5°C	—	—	—	0	—	—	6.5	—	—	—	—	—
40.0°C	5.5	8.4	10.0	0	—	11.9	5.3	12.5	18.6	42.2	—	7.9
Fatty acids composition (%)												
12:0	0.1	0.1	—	0.2	0.1	0.2	0.2	0.2	0.3	0.1	0.1	0.1
14:0	1.0	1.0	—	1.2	1.0	1.2	1.2	1.3	1.5	1.3	0.9	1.2
16:0	43.0	43.0	—	42.1	36.6	42.1	42.1	51.1	52.0	62.5	50.4	51.0
16:1	0.1	0.1	—	0.1	0	0	0.1	0.1	—	—	—	—
18:0	4.8	6.7	—	4.1	4.0	7.5	4.1	4.6	4.9	5.3	6.6	5.6
18:1	39.4	44.5	—	40.7	44.8	47.7	40.7	34.1	33.3	24.4	35.8	34.0
18:2	11.0	4.0	—	11.1	12.6	1.0	11.1	8.2	7.6	6.0	5.5	7.4
18:3	0.2	0.2	—	0.2	0.2	0	0.1	0.1	0.1	—	0.1	0.1
20:0	0.4	0.4	—	0.3	0.7	0.3	0.3	0.3	0.3	0.4	0.6	0.6
Total saturated acids (%)	49.3	51.2	—	47.9	42.4	51.3	42.4	57.5	59.0	69.6	58.6	58.5
Total monounsaturated acids (%)	39.5	44.6	—	40.8	44.8	47.7	44.8	34.2	33.3	24.4	35.8	34.0
Total polyunsaturated acids (%)	11.2	4.2	—	11.3	12.8	1.0	12.8	8.3	7.7	6.0	5.6	7.5
Polyunsaturated to saturated ratio	0.23	0.08	—	0.24	0.29	0.02	0.29	0.14	0.13	0.08	0.09	0.13

PO = palm oil   POs = stearin   H = hydrogenated   x1 = single fractionated
POo = olein   PMF = palm mid-fraction   Int. = interesterified   x2 = double fractionated

The changes in the level of palm oil in specific applications is a reflection of economic considerations.

Major applications of palm oil and its products are discussed in the following sections.

5.3.1 Oils and fats for cooking

This constitutes the largest usage of edible oils and fats. It is estimated that during 1981 about 12–16 million t of oils and fats were used for cooking human foods — more than 35 per cent of the total world consumption of oils and fats in human foods. More than 60 per cent of palm oil and olein exported by the producing countries is used by the importing countries for deep or shallow frying of foods and gravy mixing.

Fats and oils are used domestically for deep frying, but find wider use in catering and food manufacturing industries where they are used for a large variety of foods. In this application the oil/fat is used as a means of heat transfer and is used repeatedly until it has deteriorated to an unacceptable level. Its stability both in use and storage is therefore important. Oils and fats with similar shelf life may differ in their rate of deterioration at high temperature. Accumulation of double bonds in oils and fats increases their reactivity towards oxygen and as such materials with a low linoleic and linolenic acids content are used for this application, especially where the turnover rate is low. Liquid vegetable oils having a high linolenic acid content have to be partially hydrogenated to reduce their linolenic acid content to an acceptable level.

Palm oil and its modified products have excellent oxidative and frying stability due to their composition and to the presence of natural antioxidants. These are therefore viewed, especially in Europe, as the first choice for industrial frying so long as they are competitive in price with other oils and fats. Palm oil, olein, hydrogenated palm oil and blends of palm oil and stearin are widely used for this application in a number of countries. In the United States olein on its own or in blends with cottonseed oil is used for industrial frying of snacks. Palm oil, hydrogenated palm oil or blends of palm oil and other hydrogenated vegetable oils are used for industrial preparation of french fries. A number of countries in South-East Asia use palm oil and blends of palm oil and stearin for instant noodle frying. For organoleptic acceptance solids at 35 and 40 °C of such blends are kept to 15 and 10 per cent respectively.

In shallow frying, oil or fat forms an integral part of the food to be consumed; therefore, taste and flavour of the food after frying are of great importance. Health claims based on the concept of polyunsaturated fatty acids also play a role. In markets where an oil is required to remain clear on the shelf, especially in temperate countries, and also where high polyunsaturated fatty acid content is required, the usage of palm olein is limited to low levels. In some countries, specially in Japan, these problems have been overcome by blending single or double fractionated palm olein with liquid vegetable oils having high polyunsaturated acid content.

Plastic cooking fats such as cooking shortenings, palm oil, hydrogenated palm oil and palm stearin are widely used in a number of countries. In many EEC and Middle Eastern countries products containing up to 100 per cent palm oil and/or its products are being marketed. In some of these countries such products are also made from blends containing palm oil or hydrogenated palm oil or palm stearin and other hydrogenated vegetable oils.

Vanaspati, a granular, semisolid, general-purpose cooking fat, is produced and used in the Indo-Pakistan subcontinent and in some Middle Eastern countries as a major cooking medium. This fat is a single component or a blend of two or more hydrogenated oils that are slowly cooled in containers to achieve a granular consistency. In the subcontinent, depending on the availability and relative prices of different oils, vanaspati products may contain up to 70–80 per cent palm oil or hydrogenated palm oil. In some Middle Eastern countries products based on 100 per cent palm oil or blends of palm oil and stearin are also being made and marketed as vanaspati.

The use of palm oil and its products in vanaspati and plastic cooking fats has a distinct advantage and, in countries using semisolid fat as a main source of edible fats, the import of palm oil and its fractions has increased manyfold during the last few years. In India during 1983 more than 200 000 t of RBD palm oil and 300 000 t of RBD olein were imported and used as cooking fat and liquid cooking oil respectively without further processing.

5.3.2 Fats for bakery products

In the bakery trade the term 'shortening' is used to describe these fats. Initially the term was referred to lard used for pastry and bread making. Later it was extended to include cake making, frying and also creaming fats. Shortenings are anhydrous mixtures of liquid oils and fats having smooth and plastic consistency. An exception to this are liquid shortenings based mostly on liquid oils. The desirable plastic consistency is achieved by quick chilling and crystallization of the fat/blend in a scraped surface heat exchanger and a crystallizer.

Shortenings are formulated and produced to satisfy specific markets or user requirements and therefore show a wide variation in their physical and functional properties. The factors generally taken into account are type and recipe of the baked products, type of the process involved and whether the users are domestic or industrial. Shortenings for domestic use are of the 'general-purpose' type and are formulated not only for making cakes, pastry and other baked products but also for creaming and frying.

Apart from shortenings, margarine is also used for making baked products and in most applications are interchangeable. Margarines, like shortenings, are also tailor-made for specific applications.

About 4 million t of edible oils and fats are used for making shortenings and compound cooking fats in the world. During 1980 about 520 000 t of palm oil products were used in the consuming countries for making these fats — about 13 per cent of the total world consumption of edible oils and fats in these applications[5]. Of the total palm oil consumption of over 5 million t, about 11 per cent was used for making shortening during this year, mostly in the developed countries having a well-established and advanced bakery industry, especially in Western Europe, the United States, USSR and Japan.

Palm oil and its products are very useful ingredients for making plastic shortenings and very large amounts of these products are used in their formulation. This is the second largest usage of palm oil products.

Based on their applications, these fats can be divided into the following types:
1. Bread dough fat
2. Biscuit/short pastry dough fat
3. Puff pastry fat/margarine
4. Cake fats/margarine

Both plastic and liquid shortenings are used in bread and other yeast-raised baked products. The amount of shortening used in these products is small. In a number of countries texturized palm oil or palm oil/palm stearin blends are used. Sometimes an emulsifier such as glycerol monostearate is added to such shortenings to improve the bread volume and crumb texture. Hard palm stearin (m.p. 55–85 °C) is also used as a dough improver.

In biscuit and short pastry systems, the main function of the fat is to coat flour particles, especially protein, in order to make them impermeable to water when liquid ingredients are mixed in. In this way a shorter texture is obtained. For these applications, consistency and spreadability of fat are very important. A too stiff shortening is difficult to spread evenly onto the dry ingredients, whereas a too soft shortening adversely affects the machinability of the dough. Here again shortening based on palm oil, hydrogenated palm oil or blends of palm oil and palm stearin are widely used. In a number of countries blends of palm oil products and hydrogenated liquid oils or lauric oils are also used. Palm-oil-based products do not require any hydrogenation or require only minimum hydrogenation to obtain the desired consistency and shortenings based on them are less expensive to make.

Fats used for this application must have a high resistance to oxidative stability. Because of its chemical nature, palm oil products have good oxidative stability and are most suited for this application.

In puff pastry two types of fats are required, the dough fat and a layering fat. The dough fat is similar to biscuit/short pastry fat. The layering fat, generally a margarine, must be very plastic and tough since these are the characteristics required for the development of the fine even-layered structure of puff pastry. Margarines which are too soft or which soften too readily on working are liable to break down during rolling. They tend to be absorbed into the dough, resulting in a coalescence of the layer of poor lift-off on baking. Margarines that are hard or brittle may break through the dough layers, causing poor or uneven baking.

Traditional puff pastry fats have a slip melting point in excess of 44 °C. They are very tolerant towards misuse and give extremly high lift-off because they are very plastic. Unfortunately they impart a noticeable waxy aftertaste because they do not melt completely in the mouth. Intermediate melting point puff pastry fats melting at 39 °C or below do not give the waxy aftertaste. These are generally more sensitive and, therefore, need more care in handling.

Palm oil, hydrogenated palm oil and palm stearin are excellent ingredients and are widely used in the formulation of those fats. In some countries up to 90 per cent of palm oil products are incorporated in these fats.

The fats used in making cakes generally fall into the following three groups:
1. Plastic shortening
2. High ratio cake shortening
3. Liquid shortening

Plastic and high ratio cake shortenings generally have flat solid fat content profiles. These have 15–25 per cent solids at 20 °C and melt above 38 °C. However, even if the solid fat content profile and melting point of two cake shortenings are similar, performance in the cake can still be affected by their structural and compositional differences.

Palm oil and its products have the tendency to crystallize in β' crystalline form and to perform effectively in cakes. Shortening should crystallize in this form. Because of its crystalline nature and plastic consistency without hydrogenation, palm oil products are widely used in this application in the United States, the United Kingdom and other European countries. The amount of palm oil products in plastic cake shortening generally varies from 30 to 40 per cent. Some manufacturers, however, are able to incorporate as much as 80 per cent of palm oil and its products in their formulation.

5.3.3 Oils/fats for table margarine

Table margarines generally fall into the following three groups:
1. Packet margarine
2. Soft tub margarine
3. PUFA (polyunsaturated fatty acids) margarine

The fat component of margarine, representing 80–85 per cent of the product, largely determines the physical properties of the product. The physical properties of firmness and spreadability of margarine are mainly related to the proportion of solid fat and liquid oil at a given temperature. The fat phase used in all three types of margarines is usually a mixture of solid fat and liquid oil chosen to obtain the desired solids content at a range of temperatures. Table margarines in packets have enough solids between 10 and 20 °C to be easily spreadable at room temperatures, but are firm enough to retain their shape in the packet. Tub margarines are appreciably softer and generally have enough solids between 5 and 10 °C to spread straight from the refrigerator. PUFA margarines are also soft and spreadable straight from the refrigerator. To meet the claim of high polyunsaturated fatty acids (PUFA) content, the liquid oil portion of the fat phase consists of a high PUFA vegetable oil, such as sunflower, corn oil, partially hydrogenated soyabean oil, etc.

Palm oil by virtue of its triglyceride composition has a tendency to crystallize. In the β′ form required to give plastic consistency to the product. It also has a higher solids content than most other vegetable oils — quite close to that of margarine fat blends — and also shows a wide plastic range. This makes it a particularly valuable raw material in countries where hardening capacity is absent or limited. A number of tub margarine products based on 100 per cent palm oil are being made and marketed in Australia and in South-East Asian countries. Such products are also being made in EEC countries for export to some African and Middle Eastern countries with tropical climates. Higher amounts of palm oil in packet margarine formulation adversely affect the packetability of the products because of its slow crystallizing behaviour. This problem, however, has been overcome by some manufacturers by deeper chilling of the emulsion and by using a larger B-unit.

In spite of these advantages, levels of inclusion of palm oil and its products in the three types of margarine, especially those formulated for temperate climates, are limited because of their slow crystallizing and melting behaviour due to their peculiar triglycerides composition. This applies more strongly to premium brand soft tub and PUFA margarines which must spread at low temperatures and have a good oral response at high temperatures. Higher amounts of palm oil products generally impart a rather slow or thick oral melting performance. Hardened palm oil, because of the trans isomers, has a much higher solid content than palm oil. This makes it an ideal hard stock for fat blending, but limits its use to around 20–30 per cent in many West European margarine fat blends.

Palm olein also does not exhibit quick melting in the mouth as required, especially in premium brand margarine. In its contribution to spreadability and oral response, palm olein does not differ greatly from palm oil. Palm oil, palm olein (single fractionated) and palm stearin all have a solid fat content higher than 45 per cent at refrigerator temperature. The margarine based on higher levels of these products would not have the spreadability requirements of soft tub margarine at 5 °C. These products, therefore, have to be blended with partially hydrogenated or unhydrogenated liquid vegetable oils to obtain the required consistency. This reduces the level of inclusion of these products in the blends.

Interesterification of palm oil products increases the acceptable level of inclusion of these products in all the three types of formulations. Randomization of palm oil or palm olein on its own results in an increase of the trisaturated types of glycerides with a corresponding increase in melting point. Randomization with lauric oils increases the number of triglyceride types and the content of trisaturated triglycerides, but the presence of medium-chain fatty acids reduces the melting point and gives sharper melting in the mouth. By selecting suitable component ratios a wide range of fats is produced which have solids contents ideally suitable for use at higher levels. In some West European, Australian and South-East Asian countries margarines, containing up to 80 per cent of such interesterified blends in their fat phase are being produced.

Interesterification with lauric oil is one of the means of utilizing low-priced palm stearin in premium margarine fat blend. The interesterification reduces the solid content of randomized blends, especially at 20 °C and above, to levels suitable for margarine manufacture and the oral melting properties are correspondingly improved. By interesterifying palm oil products, particularly palm stearin, with high PUFA vegetable oils, fat blends suitable for PUFA-type margarine are also obtained.

It is estimated that about 6.3 million t of edible oils and fats were used in making margarine during 1981[5]. Of this amount only 380 000 of palm oil products were used in their formulation, which is about 8.3 per cent of the total consumption of palm oil products during that year. The use of palm oil products in European table margarines is well established. Very small amount of these products, however, are used in US margarines. During 1981–2 only 2300 t of palm oil products was used in the United States in margarine formulation, which was much less than 1 per cent of the total of 1.17 million t of oils and fats used for making margarine during this year[6].

In countries where the addition of artificial colouring is not permitted, crude or lightly processed red palm oil is added to the fat blend to give colour to the margarine.

5.3.4 Confectionary fats

Fats for confectionary products are required to have reasonably high solid fat contents between 20 and 30 °C. Therefore, the natural fats having a higher solid fat content and vegetable oils modified by the process of hydrogenation, interesterification or fractionation to have high solid fat contents at these temperatures are used in such products.

Confectionary products having fat as one of their components fall into the following two groups:
1. Sugar-based products
2. Chocolate-based products

In sugar-based products such as caramels, fudges, nougats, etc., the fats used were traditionally cocoa butter and dairy butter. Due to the escalation in the prices of these fats, they have been replaced with coconut and palm kernel oil, hydrogenated to the correct melting point and with a suitable SFC profile. These fats are now widely used in these products. However, to use these fats, great care must be taken with the hygenic conditions of the ingredients used and it is necessary to check for contamination after processing to avoid lipolytic rancidity. Because of these problems some manufacturers prefer to use fats based on non-lauric oils. Palm mid-fractions and hydrogenated palm olein having an appropriate solid fat content profile have been marketed by some European and Canadian refiners for such applications. However, the use of these products is not widespread and fats based on palm kernel oil and coconut oil, because of their excellent oral melting properties, still command the major share of this application.

In most countries the fat phase in chocolate products should consist by law of only two components, viz. cocoa butter from cocoa beans and milk fat. However, in the United Kingdom, Ireland and Denmark, the addition of 5 per cent vegetable fat (foreign fat) other than cocoa butter and milk fat is allowed in chocolate products. Some East European countries even allow up to 7 per cent of foreign fat in these products. These non-cocoa-butter vegetable fats are known as cocoa butter extenders (CBE). In order to be of use they must be totally compatible with the fat system they are used with at the level concerned.

Almost all CBE having high compatibility with cocoa butter are formulated from fats and their fractions containing triglycerides similar to the ones present in cocoa butter. Palm oil is rich in one of such triglycerides, i.e. POP (2-oleodipalmitin). Palm oil is fractionated to obtain a mid-fraction enriched with these triglycerides. It is estimated that about 150–200 thousand t of palm oil is fractionated annually to produce about 45 000–60 000 PMF for this application. Usually a solvent fractionation process is used to produce fractions of the quality suitable for this application. A good PMF is used directly as a CBE in plain chocolate and also in milk fat containing 1–2 per cent milk fat. To produce CBE with increased compatibility with cocoa butter and milk fat, PMF is mixed with shea stearin and/or illipe fat.

Palm olein hydrogenated on its own or in blends with shea olein to a steep solid fat content profile and melting point of 37–38 °C is also used for making compound coatings, pastel coatings and other snack coatings. Its use, however, is limited.

5.3.5 Miscellaneous

Palm oil and olein, because of their oxidative stability, are widely used in filled milk and liquid coffee whitener. Palm oil on its own or blended with palm kernel oil is also used as a fat phase in non-dairy ice cream for chocolate coating for ice cream lollies and bricks, and in sandwich cream. Hydrogenated palm oil and palm stearin are also used in dried soups and powder mixes.

5.4 References

1. MES (UK) Limited, 'The UK market for palm oil', Report submitted to the Palm Oil Research Institute of Malaysia, Kuala Lumpur, Malaysia (1982).
2. IMES (UK) Limited, 'The UK market for palm oil', Report submitted to the Palm Oil Research Institute of Malaysia, Kuala Lumpur, Malaysia (1982).
3. B.K. Tan and C.H. Oh Flingoh, 'Oleins and stearins from Malaysian palm oil: Chemical and physical characteristics', PORIM Technology 4, May 1981. Published by Palm Oil Research Insitute of Malaysia, Kuala Lumpur, Malaysia (1981).
4. J. Rourke, 'Use of palm oil in the edible oil business', Paper presented at the 50th Fall Meeting of AOCS, from 26 to 29 September 1976 in Chicago, USA (1976).
5. Y. Basiron and J. Abdullah, 'Palm oil: World supply and demand', Paper prepared for Palm Oil Research Institute of Malaysia, Kuala Lumpur, Malaysia (1982).
6. Anon., 'Fats and oils situation and outlook', Commodity Year Book, 1982, USDA/ERS (1982).

Animal feed

R.I. Hutagalung

5.5	**Introduction**	84
5.6	**Palm oil and its byproducts**	84
5.7	**Palm press fibre and palm empty fruit bunch**	87
5.8	**Palm oil mill effluent**	88
5.9	**References**	89

5.5 Introduction

Since 1966, Malaysia has become the world's largest palm oil producer, currently accounting for more than 65 per cent of the global trade. Presently, oil palm cultivation covers nearly 1.4 million ha, producing nearly 4.0 million t of palm oil (PO), 0.5 million t of palm-kernel cake (PKC), 0.2 million t of dried palm oil mill effluent (POME) (equivalent to 4.2 million t of fresh POME), 2.5 million t of palm press fibre (PPF) and 12.5 million t of palm empty fruit bunch (PEFB). The corresponding estimated production in Asia and the world was respectively (in million tonnes): 5.0 and 6.5 for PO, 0.6 and 0.8 for PKC, 0.25 and 0.35 for dried POME, 2.9 and 4.0 for PPF and 14.5 and 20.0 for PEFB.

Oil palm products, byproducts and residues that either have been utilized or have considerable potential as animal feeds include crude PO, palm olein, palm stearin, PKC, POME and PPF. The potential of oil palm residues such as palm-kernel shells, palm fronds and trunks as animal feeds has not been studied, possibly due to the constraints in collection, processing and the highly lignified nature of these residues. Among these, PO and PKC have been added most frequently into the diets of farm animals. The use of POME and PPF is still limited, mainly under experimental or pilot trials, but the future use of these residues holds promise when knowledge on processing technology for POME and on the fibre characteristics of PPF in relation to the digestive function in the ruminants is fully understood.

Information on the utilization and state of processing palm products, byproducts and residues as animal feeds has been documented[1–13]. Further discussion of the use of palm-kernel cake is excluded because this review is confined to palm oil.

5.6 Palm oil and its byproducts

Crude PO and its major refinery byproducts, olein and stearin, have been routinely used in pig and poultry feeds as sources of energy. Crude PO is yellowish-red in colour due to the presence of carotenoids comprising mainly of α- and β-carotenes

Table 5.4 Chemical composition of crude palm oil, palm olein and palm stearin[17]

Item	Palm oil	Palm olein	Palm stearin
Free fatty acid (%)	2.5–6.5 (4.24)	2.8–5.0	2.0–2.9
Fatty acids (%)			
Lauric (12:0)	0.1–1.0 (0.23)	0.1–1.1 (0.19)	0.1–0.6 (0.1)
Myristic (14:0)	0.9–1.5 (1.09)	0.9–1.4 (1.04)	1.1–1.9 (1.2)
Palmitic (16:0)	41.8–46.8 (44.02)	37.9–41.7 (39.76)	47.2–73.8 (52.6)
Palmitoleic (16:1)	0.1–0.3 (0.12)	0.1–0.4 (0.17)	<0.05–0.2 (0.1)
Stearic (18:0)	3.7–5.6 (4.54)	4.0–4.8 (4.39)	4.4–5.6 (4.9)
Oleic (18:1)	37.3–40.8 (39.15)	40.7–43.9 (42.49)	15.6–37.0 (32.4)
Linoleic (18:2)	9.1–11.0 (10.12)	10.4–13.4 (11.17)	3.2–9.8 (8.1)
Linolenic (18:3)	<0.05–0.6 (0.37)	0.1–0.6 (0.40)	0.1–0.6 (0.1)
Arachidic (20:0)	0.2–0.7 (0.38)	0.2–0.5 (0.37)	0.1–0.6 (0.4)
Iodine value	51.0–55.3 (53.35)	56.1–60.6 (58.03)	21.6–49.4 (44.1)
Melting point (°C)	32.3–39.0 (35.96)	19.4 –23.5 (21.57)	44.5–56.2 (49.1)
Peroxide value	0.40–8.79 (1.28)	—	—
Vitamins			
Carotene (ppm)	460–819 (650)	476–647 (565)	125–379 (248)
Tocopherol (ppm)	635–100 (792)	796–994 (865)	251–530 (37.2)
Energy, metabolizable (MJ/kg)			
Poultry	36.4–37.2	35.6–36.4	36.0–36.8
Swine	36.0–36.8	35.1–36.0	35.6–36.4
Digestibility (%)	85–96	84–92	88–94
Iron (ppm)	3.45–9.60 (6.16)	0.6–4.3	16.1–21.0
Copper (ppm)	0.00–0.51 (0.04)	0.05–0.06	0.06–0.07
Phosphorus (ppm)	13.1–38.2 (22.5)	—	—

Sources: B. Bek-Nielsen and S. Krishnan (1976); Hutagalung *et al.* (1981); Tan *et al.* (1983). Int. Dev. Palm Oil Proc. Malays. Int. Symp. Palm Oil Process Mark 291–297

and traces of γ-carotene, as well as other xanthophylls[14–17]. Crude PO is semisolid at room temperature, consisting mainly of palmitic (~44 per cent), oleic (~39 per cent) and linoleic (~10 per cent) acids (Table 5.4). The presence of a large amount of linoleic acid in PO renders it more suitable for pigs and poultry as it is required for growth and production. Palm olein and stearin have 40 and 53, 42 and 32, and 11 and 8 per cent for palmitic, oleic and linoleic acids respectively. The melting point of PO, olein and stearin average 36, 22 and 49 °C respectively.

In Malaysia, PO is usually added as an energy supplement in pig and poultry diets to substitute for costly imported animal fats (tallow and lard), and the inclusion of up to 3 per cent of PO in the diets is currently practised. Studies in pigs and poultry on the use of PO as an energy source have shown superior responses in rate of gain and feed efficiency, compared to fats of animal origin[18, 19].

Ogunmodede and Ogunlela[20] compared the effects of groundnut oil, melon seed oil and palm oil supplementation on the performance of pullets (White Plymouth Rock and Rhode Island Red). Supplementation at 6 and 10 per cent of palm oil markedly improved calcium absorption, while melon seed oil and groundnut oil were effective only at the 10 per cent level. Egg production was improved only on White Plymouth Rock pullets. A low cholesterol content of egg yolk was recorded from layers fed an oil palm supplemented diet. In Nigeria, Oluyemi and Okunuga[21] demonstrated that palm oil supplementation (0, 2.5, 5, 7.5, 10 per cent) to layer diets for 8 months improved feed efficiency, fertility (5–7.5 per cent) and hatchability (2.5–7.5 per cent) of laying hens, indicating the contributive effect of energy and essential fatty acids from the palm oil.

The increasing demand for high-energy (density) diets, particularly for broiler chickens and young pigs, in Malaysia has intensified use of palm oil to supplement energy. It has been a common practice to add palm oil at 2–6 per cent in the diets, depending on the level of energy desired and type of feeding stuffs used in the diet. Salmon and O'Neil[22] compared the effectiveness of palm oil and rapeseed oil supplementation at 0, 2 and 11.4 per cent to turkey broilers. A marked improvement in growth, feed efficiency and carcass quality was observed at 2 per cent oil supplementation. A further increase up to 11.4 per cent was effective for palm oil but not for rapeseed oil. In an attempt to improve the utilization of palm kernel cake (PKC) in the broiler diet, palm oil at various levels (0, 3, 6, 9, 12 per cent) was included in 20% PKC-based diets[23]. Weight gain and feed efficiency were superior in table birds fed 9 and 12 per cent palm oil and maize–soyabean meal control diet than those without or with 3 and 6 per cent added palm oil. There were, however, no appreciable differences in carcass quality and mineral composition among the palm oil added diets.

It has been recognized that fat or oil addition in the diet could improve the efficiency of energy utilization by 10–15 per cent in broiler chickens. Palm olein and stearin are as effective as palm oil, but the extent of their use is dictated by price[4, 19, 24].

Studies in Malaysia and Nigeria on caloric to protein ratio requirements of pigs indicate that increase in the level of palm oil in the diet (beyond the 5 per cent level) increases fat content and backfat measurements[18, 19, 25–27]. However, other reports indicate that addition of palm oil at 2–10 per cent[28] and at 0–30 per cent[29] showed no marked effect on feed efficiency and carcass quality. Palm oil is generally added to carbohydrate(cassava and sago)-based diets to reduce dustiness and improve texture and intake of the diet[30–33].

The use of fat or oil in the diet of ruminants is relatively less frequent than for non-ruminants, simply due to their preference for carbohydrates as sources of energy rather than fats. Hence the information on palm oil addition in ruminant concentrate is limited. Morever, locally available feedstuffs such as palm-kernel cake, palm oil mill effluent, cassava and sago can be readily utilized by the ruminants[6]. Dietary supplementation of palm oil at the rate of 2–8 per cent has been shown to increase milk fat content[34]. On the other hand, addition of palm oil (8 per cent) to the diet of sheep reduced dry matter and protein digestibility[1]. It can be concluded that the level of palm oil, palm olein and palm stearin supplementation is governed by economic factors. Results from digestibility and metabolizable energy studies of palm oil in pigs and poultry showed the range of 85–96 per cent for digestibility and 36.0–36.8 MJ/kg for mill effluent[35]. Under a practical feeding system, palm oil was added at 2–6 per cent in the diets of pigs and poultry, depending on the stage of development and productivity of animals.

5.7 Palm press fibre and palm empty fruit bunch

Palm press fibre (PPF) and palm empty fruit bunch (PEFB) are the fibrous residues separated from the kernels during the palm oil extraction, accounting for 10–14 and 55–60 per cent of the fresh fruit bunch respectively. PPF can be classified as lignocellulose material due to its high fibre (ADF 40.7 per cent; NDF 75.4 per cent) and lignin (21.5 per cent) content, but is low in protein (7.9 per cent). The large quantity of PPF available in Malaysia and other oil palm growing countries offers the opportunity as a potential roughage material for feeding ruminants. However, the poor intake and digestibility of PPF make it unattractive as a ruminant feedstock.

A number of studies have been made to evaluate and improve the nutritional value of PPF. Chemical treatment with various alkali solutions (6% NaOH or 8% $Ca(OH)_2$) had no significant effect on nutrient digestibility of PPF in sheep, apparently due to an inappropriate method of treatment[2], compared with more recent attempts of treating PPF with NaOH, NH_4OH or urea, where a higher nutrient digestibility was recorded. Low *in sacco* dry matter disappearance (ISDMD) rates of 26.3 and 30.6 per cent for untreated PPF and 28.2 and 34.3 per cent respectively for urea-treated (5 per cent urea; 30 days incubation) PPF upon 48 h incubation in the rumen of cattle and buffalo was reported[36]. A similar, but more marked, response was demonstrated when PPF was treated with NH_4OH (6 and 50 per cent moisture; 24 d incubation); the improvement in ISDMD rates was from 27.5 to 35.8 per cent and 33.0 and 39.9 per cent when incubated in the rumen of cattle and buffalo respectively for 48 h (Vijchulata *et al.*, 1985)[36]. The interaction of ammonia or ammonium hydroxide with the lignocellulose structure of fibre may account for the improvement in rumen digestibility of treated PPF. The digestibility of PPF by the rumen liquor of buffalo was greater than that of cattle for both untreated and treated PPF, showing the more effective dry matter and cellulose digestion by buffalo than cattle[36]. Results from *in vivo* digestibility and performance studies in cattle and sheep fed ammonia-treated PPF (4% NH_4OH; 21 d incubation) confirmed the improved effects in intake and digestibility of treated PPF at *in vitro* scale[6, 7].

Supplementation of small amounts of good-quality protein (fish meal) and energy (rice bran) to PPF-based diets in the form of byproducts has been shown to be effective in cattle (palm-kernel cake[7, 8, 11] and in buffalo (360 g fish meal[37]). The improvement has been due to an increase in dry matter intake and rumen volatile fatty acid contents. Earlier work on the supplementation of molasses, urea and cassava to the feeds of cattle and buffalo supported a satisfactory performance in intake, growth rate and feed conversion[38, 39].

Results from a comparative study on the rumen functions of cattle and buffalo given either grass or PPF showed no appreciable difference in the ureolytic activity on the wall of the reticulo-rumen. Both buffaloes and cattle on grass-based diet showed similar areas with high ureolytic activity, including the reticulo-rumen areas at ventral and dorsal surfaces of the caudal pillar, tip of the cranial pillar and the dorsal surface of the ventral sac. However, the overall urease activity in buffaloes fed PPF appeared to be greater than cattle fed the same diet, presumably due to the increase in activity at the fundic region of the bufallo's abomasum. The rumen ammonia level seemed to exert a feedback control on the urease activity of rumen bacteria; a more pronounced activity in buffalo was accompanied by higher rumen ammonia nitrogen than in cattle[40].

The paucity of information on PEFB is reflected by the neglected interest in its use as a roughage source. It is frequently left to rot at the mill although on rare occasions it is burnt for fertilizer use. Paper has been produced on an experimental basis from PEBF by a chemical process. There is no information available on the feeding value of PEBF except that, at the *in vitro* stage, dry matter digestibility was observed to be lower than PPF. In an unpublished study, the author found that the dry matter disappearance of PEFB in the rumen of cattle and buffalo was 32 and 37 per cent respectively at 48 h.

5.8 Palm oil mill effluent

Palm oil mill effluent (POME) refers to the collective term for liquid wastes of discharge from the final stages of palm oil production at the mill. The 5 per cent total solids of POME include dirt, residual oil and suspended soil. Various processing methods and utilization in animals have been well documented [2, 9, 13]. Results of chemical composition of POME collected from various oil palm mills showed a large variation in the products, especially in ash (15.1–48.1 per cent), fibre (11.4–24.3 per cent) and fat (11.6–24.3 per cent) contents. High ash and fibre contents rendered POME less available for non-ruminant feed (Table 5.5). Similarly, high fat and ash contents of POME resulted in its poor utilization by the ruminants, particularly at the high rate of inclusion in concentrate feeds. Digestibility and availability of amino acids in pigs and poultry were low, particularly for lysine (8.3 per cent) and methionine (22.1 per cent)[2]. Feeding trials in pigs and poultry have shown that POME can be optimally included up to 15 per cent for broilers and up to 10 per cent for layers and pigs. In ruminants, the optimal rate of inclusion of POME ranged from 20 to 40 per cent for cattle and buffalo and 15 to 30 per cent for goats and sheep.

Although the evidence to date suggests that POME has potential as a valuable animal feed, mainly supportive to conventional energy and protein sources, there are several problems that need serious consideration, notably: (1) the limitation in using unprocessed POME *in situ* owing to transportation and short shelf life, (2) low palatability, (3) high ash content and toxic minerals, (4) large variations in ash, protein and residual oil content, (5) overheating of the dried form, reducing

Table 5.5 Chemical composition of POME, PPF and PEFB[5–7, 9, 13]

Item	POME	PPF	PEFB
Dry matter (%)	93.3	72.5	75.0
DM digestible (%)	70.2	45.2	—
Ash (%)	18.5	7.9	9.3
Crude fibre (%)	17.9	40.7	43.8
Ether extract (%)	13.0	9.4	7.1
N-free extract (%)	33.3	6.6	5.0
Protein (Nx6.25) (%)	10.6	7.9	9.8
Digestible protein (%)	6.1	3.8	—
Energy (MJ/kg)			
Gross	—	—	—
Poultry ME	7.7	—	—
Swine ME	9.2	—	—
Cattle ME	9.8	7.3	5.5
TDN (%) (ruminant)	68.0	48.0	37.0
Calcium (%)	0.38	0.26	0.21
Phosphorus (%)	1.34	0.09	0.05
Magnesium (%)	0.60	0.10	—
Copper (mg/kg)	55.0	16.0	—
Iron (mg/kg)	8700	1800	—
Manganese (mg/kg)	120.0	46.0	—
Zinc (mg/kg)	49.0	20.0	—
NDF (%)	—	75.4	78.6
ADF (%)	—	52.8	45.3
Lignin (%)	—	21.5	25.0
Hemicellulose (%)	—	22.6	—
Cellulose (%)	—	31.3	—
DMI (%) (ruminant)	1.70	1.70	1.35
Inclusive rate (%) (monogastric)	10.0	—	—
Shell (%)	—	—	—

POME = palm oil mill effluent
PPF   = palm press fibre
PEFB = palm empty fruit bunch

the availability of nutrients, and (6) insufficient data on large-scale feeding trials and appropriate feeding systems for farm animals to augment its use by the livestock industry[9].

5.9 References

1. C. Devendra, 'Utilization of feedstuffs from the oil palm', in *Feedingstuffs for Livestock in South East Asia*, ed. by C. Devendra and R.I. Hutagalung, MSAP, Kuala Lumpur (1978), pp. 116–39.
2. C. Devendra, S.W. Yeong and H.K. Ong, 'The potential value of palm oil mill effluent (POME) as a feed source for farm animals in Malaysia', in *Proceedings of National Workshop on Oil Palm By-Product Utilization*, PORIM, Kuala Lumpur (1983), pp. 63–75.
3. R.I. Hutagalung, 'Non-traditional feedingstuffs for livestock', in *Feedingstuffs for Livestock in South East Asia*, ed. by C. Devendra and R.I. Hutagalung, MSAP, Kuala Lumpur (1978), pp. 259–88.
4. R.I. Hutagalung, 'The use of tree crops and their by-products for intensive animal production', in

Intensive Animal Production in Developing Countries, ed. by P. Smith and R.G. Gunn, Edinburgh (1981), pp. 151–84.

5. R.I. Hutagalung, 'Chemical composition and nutritive value of palm kernel cake and palm press fibre', in *Recent Advances in Animal Nutrition in Australia 1983*, ed. by D.J. Farrell and P. Vohra, University of New England Publishing Unit, Armidale (1983), p. 4A.

6. R.I. Hutagalung, 'Nutrient availability and utilization of unconventional feedstuffs used in tropical regions', in *Proceedings of Feeding Systems of Animals in Temperata Areas*, 2–3 May 1985, Seoul, Korea, KFIC/INFIC/AAAP (1985), pp. 326–37.

7. R.I. Hutagalung and M.D. Mahyuddin, 'Nutritive value and feeding systems on palm kernel cake and palm press fibre for ruminants', in *Proceedings of 3rd AAAP Animal Science Congress*, 6–10 May 1985, Seoul, Korea, Vol. 2 (1985), 983–5.

8. R.I. Hutagalung and M.D. Mahyuddin, 'Oil palm by-products as feeds for animals with special reference to ruminants', in *National Symposium on Oil Palm By-products for Agro-Based Industry*, 5–6 November 1985, Kuala Lumpur, PORIM (1985), pp. 40–1 (synopsis).

9. R.I. Hutagalung, M. Mahyuddin and S. Jalaludin, 'Feeds for farm animals from the oil palm', in *The Oil Palm in Agriculture in the Eighties*, ed. by E. Pusparajah and P.S. Chew, Vol. 2, ISP, Kuala Lumpur (1982), pp. 609–21.

10. R.I. Hutagalung, M.D. Mahyuddin, P. Vijchulata, S. Jalaludin and J.A. Zainal, 'Nutrient availability and utilization of feedstuffs for farm animals', in *Feed Information and Animal Production*, ed. by G.E. Robards and R.G. Packham, CAB-INFIC (1983), pp. 497–506.

11. R.I. Hutagalung, M.D. Mahyuddin, B.L. Braithwaite, P. Vijchulata and S. Dass, 'Performance of cattle fed palm kernel cake and palm press fibre under intensive system', Proceedings 8th MSAP Conference on *Feeds and Feeding Systems for Livestock*, Genting Highlands, Malaysia (1984), pp. 87–91.

12. B.H. Webb, R.I. Hutagalung and S.T. Cheam, 'Palm oil waste as animal feed', in *International Developments in Palm Oil*, ed. by D.A. Earp and W. Newall, ISP, Kuala Lumpur (1976), pp. 125–45.

13. S.W. Yeong, T.K. Mukherjee and R.I. Hutagalung, 'The nutritive value of palm kernel cake as a feedstuff for poultry', in *Proceedings of Workshop on Oil Palm By-Product Utilization*, PORIM, Kuala Lumpur (1983), pp. 100–7.

14. S.Y. Chooi, H.F. Koh and K.H. Goh, 'A study of some quality aspects of crude palm oil — Interrelationships of quality characteristics of fresh crude palm oil and a proposed method for oil classification', in *Palm Oil Product Technology in the Eighties*, ed. by E. Pusparajah and M. Rajadurai, ISP, Kuala Lumpur (1983), pp. 217–30.

15. A. Gapor, M. Top and K.G. Berger, 'Effects of processing on the content and composition of tocopherols and tocotrienols in palm oil', in *Palm Oil Product Technology in the Eighties*, ed. by E. Pusparajah and M. Rajadurai, ISP, Kuala Lumpur (1983), pp. 145–56.

16. B. Jacobsberg, 'The influence of milling and storage conditions on the bleachability and keepability of palm oil', *Proceedings of Symposium on Incorporated Society of Planters*, 6–8 November 1969, Kuala Lumpur (1969), pp. 106–28.

17. B.K. Tan, F.C.H. Oh, W.L. Siew and K.G. Berger, 'Characteristics of processed palm oils', in *Palm Oil Product Technology in the Eighties*, ed. by E. Pusparajah and M. Rajadurai, ISP, Kuala Lumpur (1983), pp. 127–43.

18. G.M. Babatunde, B.L. Fetuga and V.A. Oyenuga, 'Comparative studies on the effects of feeding different types of oils at two levels on the performance and carcass characteristics of growing pigs in the tropics', *Animal Production*, **18**, 301–8 (1974).

19. R.I. Hutagalung, 'Additives other than methionine in cassava diets', in *Cassava as Animal Feed*, ed. by B. Nestel and M. Graham, IDRC-095e, Ottawa (1977), pp. 18–32.

20. B.K. Ogunmodede and B. Ogunlela, 'Utilization of palm oil, groundnut and melon seed oils by pullets', *Br. Poultry Sci.*, **12**, 187–96 (1970).

21. J.A. Oluyemei and K.O. Okunuga, 'The effects of dietary palm oil and energy on the performance of White Rock breeders in Nigeria', *Poultry Sci.*, **54**, 305–7 (1975).

22. R.E. Salmon and J.B. O'Neil, 'The effect of level and source of dietary fat on the growth, feed efficiency, grade and carcass composition of turkeys', *Poultry Sci.*, **50**, 1456–67 (1971).

23. S.W. Yeong, 'Biological utilization of palm oil by-products by chickens', unpublished dissertation, University of Malaya, Kuala Lumpur (1981), 216 pp.

24. K.K. Woo, personal communication, 1985.

25. G.M. Babatunde, B.L. Fetuga and V.A. Oyenuga, 'The effect of varying the calorie:protein ratios on the performance of pigs in the tropics', *Animal Production*, **13**, 695–702 (1971).

26. G.M. Babatunde, B.L. Fetuga, O. Odumosu and V.A. Oyenuga, 'Palm kernel meals the major

26. G.M. Babatunde, B.L. Fetuga, O. Odumosu and V.A. Oyenuga, 'Palm kernel meals the major protein concentrate in the diets of pigs in the tropics', *J. Sci. Food Agric.*, **26**, 1279–91 (1975).
27. V.F. Hew and R.I. Hutagalung, 'The utilization of tapioca root meal (*Manihot utilissima*) in swine feeding', *Malaysian Agric. Res.*, **1**, 124–30 (1972).
28. B.L. Fetuga, G.M. Babatunde and V.A. Oyenuga, 'The effect of varying the level of palm oil in a constant high protein diet on performance and carcass characteristics of the growing pigs', *E. Afr. Agric. Food J.*, **40**, 264–70 (1975).
29. C. Devendra and V.F. Hew, 'The utilisation of varying levels of dietary palm oil by growing-finishing pigs, *MARDI Res. Bull.*, **4**, 76–87 (1977).
30. V.F. Hew, 'The effect of some local carbohydrate sources such as cassava and sago on the performance and carcass characteristics of growing-finishing pigs', unpublished thesis, University of Malaya, Kuala Lumpur (1975), 177 pp.
31. V.F. Hew, 'Problematic aspects of carbohydrate sources used for pigs in Malaysia', in *Feedingstuffs for Livestock in South East Asia*, ed. by C. Devendra and R.I. Hutagalung, MSAP, Kuala Lumpur (1978), pp. 177–90.
32. K. Martin, 'Utilization of different fat sources by growing-finishing swine', unpublished thesis, University of Kentucky, Lexington, USA (1976–7).
33. S.W. Yeong and A.B. Syed Ali, 'The use of tapioca in broiler diets', *MARDI Res. Bull.*, **5**(1), 95–103 (1976).
34. J.E. Storry, H.J. Hall and V.W. Johnson, 'The effect of increasing amounts of dietary red palm oil on milk secretion', *Br. J. Nutr.*, **22**, 609–16 (1968).
35. R.I. Hutagalung, 'Utilization of feedstuffs from tree crops by poultry', *Proceedings of Second Science and Industry Seminar*, BPT, Bogor (1979), p. 166.
36. P. Vijchulata and S. Jalaludin, 'Evaluation of urea treatment on nutritive value of oil palm fibre in ruminants', in *Proceedings of 3rd AAAP Animal Science Congress*, 6–10 May 1985, Seoul, Korea, Vol. 2 (1985), 769–71.
37. Z.A. Jelan and S. Jalaludin, 'Effect of fishmeal supplementation on palm press fiber based diet in swamp buffaloes', in *Proceedings of 3rd AAAP Animal Science Congress*, 6–10 May 1985, Seoul, Korea, Vol. 2 (1985), pp. 829–31.
38. J.K. Camoens, 'Utilization of palm pressed fibre and palm kernel cake by growing dairy bulls', in *Proceedings of Seminar on Integrated Animal Planted Crops*, MSAP, Kuala Lumpur (1979), pp. 115–31.
39. R. Dalzell, 'A case study on the utilization of effluent and by-products of oil palm by cattle and buffaloes on an oil palm estate', in *Feedingstuffs for Livestock in South East Asia*, ed. by C. Devendra and R.I. Hutagalung, MSAP, Kuala Lumpur (1978), pp. 132–41.
40. N. Abdullah, M. Mahyuddin, S. Jalaludin and P. Vijchulata, 'Comparative ureolytic activity of epithelial bacteria from the GIT of swamp buffalo and cattle', Research Coordinating Meeting on *Improvement of Buffaloes*, Bangkok, Thailand (28–30 April 1985), 17 pp.

Industrial uses

R.J. de Vries

5.10	**Introduction**	92
5.11	**Fatty acid production**	94
5.11.1	Splitting	94
5.11.2	Glycerine	94
5.11.3	Distillation	94
5.11.4	Hydrogenation	95
5.11.5	Fractionation	95
5.12	**Soap making**	96
5.13	**Tin plating**	97
5.14	**Cosmetic and pharmaceutical applications**	97
5.15	**Palm methyl esters**	97
5.16	**Textile lubricants**	98

5.10 Introduction

Palm oil is relatively new to the industrial scene and initially its applications were limited because of its poor and variable quality. However, following improved quality for edible purposes palm oil of acceptable and consistent quality has become more generally available. Apart from a few established industrial applications, the oil is mainly finding application as a replacement for other triglyceride oils. Crude palm oil and RBD (refined, bleached, deodorized) palm oil are excellent raw materials for industrial purposes. However, because of their major application for edible purposes their price reacts to the international price of other edible oils such as soyabean oil, rendering them too expensive for most industrial purposes. It is mainly as a replacement for inedible tallow that the byproducts of palm oil find their industrial use (Table 5.6).

The following byproducts result from the fractionation of refining of crude palm oil for edible purposes:

1. Crude palm stearin is the solid fraction of palm oil obtained by filtration or centrifugation after the oil has been crystallized at a controlled temperature.
2. Palm fatty acid distillate or deodorizer distillate is obtained as a byproduct from the physical refining of palm oil.
3. Palm acid oil is a general term for a byproduct obtained from alkali refining of oils and fats. During alkali refining, the free fatty acids are neutralized with

Table 5.6 Fatty acid composition of tallow and palm oil

	Myristic 14:0	Palmitic 16:0	Stearic 18:0	Palmitoleic 16:1	Oleic 18:1	Linoleic 18:2
Tallow	3.0	25.0	21.5	2.5	42.0	3.0
Palm oil	1.0	43.7	4.4	—	39.9	10.3
Hydrogenated tallow	3.0	27.5	66.5	—	—	—
Hydrogenated palm oil	1.0	43.7	54.6	—	—	—

alkali and this soapstock containing some emulsified neutral oil is separated. Acidification of the soapstock gives acid oil. The main components of acid oils are free fatty acids, neutral oil and moisture.

Palm stearin is the preferred raw material, since, as a triglyceride, it yields glycerine on processing. The cost of crude palm stearin is approximately 90 per cent that of crude palm oil, whereas palm fatty acid distillate and palm acid oil cost approximately 75 per cent of the cost of crude palm oil. However, these materials are mixtures of fatty acids and contain virtually no glycerine, which is normally the most valuable product of the fatty acid separation process. In the selection of raw materials, apart from the consideration of suitability for the material to be produced, the value of resultant byproducts must be considered.

In 1984 it was estimated that Malaysian refiners produced 70 000 of crude palm stearin and 120 000 t of distillate and acid oils, which was only slightly more than that required to supply the countries' installed oleochemical capacity of 150 000 t.

As a replacement for inedible tallow these palm-derived raw materials have a number of advantages. While the international production of tallow is not increasing, planted acreage of oil palm ensures increasing availability of crude palm oil for the remainder of this century and therefore an increasing availability of the refinery byproducts useful for industrial purposes. Furthermore, palm oil and its byproducts are vegetable-derived and, provided that careful raw material selection is made together with controlled processing, they can be used in the production of those oleochemicals required for edible applications such as food emulsifiers, etc. Such oleochemicals, following specified approval, are accepted by Jewish communities as *kosher* and by Muslim communities as *halal*.

5.11 Fatty acid production

Palm oil is split to yield glycerine and fatty acid. This reaction takes place under normal handling conditions: one of the main objectives of refining an oil is to remove the fatty acid and the specification for an edible oil carries a maximum free fatty acid (FFA) requirement.

5.11.1 Splitting

The triglyceride is split as follows:

$$
\begin{array}{l}
\text{CH}_2\text{OCOR} \\
| \\
\text{CHOCOR} + 3\,\text{H}_2\text{O} \longrightarrow \\
| \\
\text{CH}_2\text{OCOR}
\end{array}
\quad
\begin{array}{l}
\text{CH}_2\text{OH} \\
| \\
\text{CHOH} \qquad + \; 3\,\text{RCOOH} \\
| \\
\text{CH}_2\text{OH}
\end{array}
$$

Triglyceride   Glycerol   Palm fatty acids

The palm oil is introduced into a splitting tower at the base, while steam is introduced at the top. The two liquids flow countercurrently at a temperature of 255 °C and a pressure of 55 bar. Palm fatty acids are taken from the top of the reactor and dilute glycerine or 'sweet water' from the base.

5.11.2 Glycerine (Glycerol)

The sweet water, which is approximately 10 per cent active, is treated to remove inorganic salts; then excess water is evaporated to yield a product with 88 per cent activity, called crude glycerine. The crude glycerine is distilled under vacuum to yield 99.5 per cent activity and is then bleached to give pharmaceutical grade glycerine.

Glycerine is used as a humectant in tobacco and cosmetics, for the production of glycerol esters such as glycerol monostearate, in food, pharmaceuticals and in explosives for the production of trinitroglycerine.

5.11.3 Distillation

The resulting fatty acids, distilled to remove water and impurities, have the specifications and chain length composition shown in Table 5.7.

Table 5.7 Distilled palm fatty acid characteristics

Titre	45–48.5 °C
Iodine number	48–56
Acid number	203–209
Saponification number	204–210
Colour 5.25 in Lovibond	FAC 1
14:0	1%
16:0	43/48%
18:0	3/5%
18:1	40%
18:2	10%

5.11.4 Hydrogenation

Distilled palm fatty acid has 40 per cent 18:1 (oleic) and 10 per cent 18:2 (linoleic) content. These two C_{18} acids may be converted to stearic acid by adding hydrogen to the double bond. Hydrogen at a pressure of 25 bar is applied to the unsaturated acids in the presence of a nickel catalyst. The degree of hydrogenation is measured and controlled by the iodine value.

5.11.5 Fractionation

Fatty acids emerge from the splitter as mixtures. Distillation will increase their overall purity and hydrogenation will alter their composition, but chain lengths are unchanged.

Distilled palm fatty acids can be fractionated to yield the values given in Table 5.8.

Table 5.8 Fractionated distilled palm fatty acid characteristics

	Stearic 90%	Oleic acid
Titre	65–69 °C	23 °C max.
Iodine number	1.0 max.	94–98
Acid number	195–201	195–203
Saponification number	196–202	197–205
Colour 5.25 in Lovibond	5Y–0.5R	10Y–1.5R
16:0	3/8%	1/2%
18:0	90/95%	7/9%
18:1	1.0%	70/74%
18:2	Trace	17/18%
18:3	Nil	1% max.

Fatty acids are used in rubber processing as lubricants and accelerators. They are also used for the production of heavy metal soaps for use in the plastics industry, metal-working lubricants, emulsion polymerization, etc. The finest candles are made from stearic acid and the use of fats and oils for this purpose dates back to prehistory. Currently, paraffin wax is predominately used for candles but the increasing cost of this petroleum-based wax makes stearic acid increasingly attractive for candle manufacture.

Apart from their direct industrial application, the majority of fatty acids is used in the production of downstream products such as long chain esters, alcohols, amines, imidazolines and heavy metal soaps. In 1980, the applications for fatty acids given in Table 5.9 were reported in the United States.

Table 5.9 Applications for fatty acids

	Percentage of total
Used directly	45
Nitrogen derivatives (amines, amides)	15
Esters	7
Metallic salts	7
Dimers and trimers	6
Ozonolysis	3
Fatty alcohols	12
Other	5

5.12 Soap making

Palm stearin is used in soap production because of its technical characteristics. Crude palm stearin is used in household and laundry soaps, while RBD palm stearin is used in the manufacture of high-quality laundry and toilet soaps. The suitability of a particular grade of palm stearin for a particular soap depends on the requirements of the consumer.

Soaps are the sodium and potassium salts of fatty acids derived from oils and fats of vegetable or animal origin. The costs of production and properties of a soap depends upon the type and the properties of the oil used, since these constitute more than 90 per cent of the basic raw materials.

When selecting a mixture of fats for soap making, one should ensure that it contains the proper ratio of saturated and unsaturated and long and short chain fatty acids to give the desired qualities of stability, solubility, ease of lathering, hardness and detergency in the finished product. Fats used in soap manufacture are coconut oil, palm-kernel oil, tallow, palm stearin and palm oil. Fats containing a high percentage of lauric and myristic acids give soaps that are readily soluble in cold water and have good foaming properties. Soaps made from soft fats and oils containing a higher percentage of unsaturated acids give soaps that are quite soluble in cold water, whereas fats such as tallow and palm stearin, which contain a high percentage of long chain saturated fatty acids, give hard soaps, requiring hot water for solubility.

For stability and pretreatment, palm stearin is superior to inedible tallow because it contains less unsaturated acids than tallow and is free of nitrogenous materials. Hence required pretreatment is simple. Palm stearin is also relatively free from odour, producing soaps that require less perfume than tallow-containing soaps.

5.13 Tin plating

During tin plating, cold-reduced steel is degreased to remove rolling lubricant, acid pickled to remove scale, rinsed in cold water and introduced into the inlet compartment of a bath through a zinc chloride flux. The steel sheet leaves from the bath of molten tin via a series of rolls immersed in an oil bath. The oil is recirculated through a jacketed reservoir to control temperature which is maintained slightly above the melting point of tin at approximately 240 °C. The oil protects the tin from oxidation, absorbs metal oxides and flux residues, maintains the tin in a molten condition as the metal sheet emerges from the metal bath and protects the newly tinned sheet from oxidation. Sawdust or bran is used to remove the thin oil film from the tinned sheet.

Crude palm oil is used in tinning because it has a composition near the optimum for the purpose, and usually contains free fatty acids that promote wetting of the sheet steel by the tin. A certain degree of heat degradation in the oil is desirable because it produces free fatty acids and possibly other compounds that assist in dissolving metallic oxides and promote uniform wetting of the metal by the oil.

Hot dipped tin plating is limited to applications that have a heavier tin layer than the usual 0.0001 in thickness produced by electroplating.

5.14 Cosmetic and pharmaceutical applications

Oils and fats are used in the formulation of cosmetic and pharmaceutical creams and ointments. Refined grades of palm oil are commonly used in these applications.

In ointments used for topical application, the oil acts as a carrier for the medicant, which must be deposited onto the skin (percutaneous absorbed). The effectiveness of palm stearin has been compared with aqueous cream, wool alcohols and soft white paraffin, containing salicylic acid, benzocaine and methyl salicylate. Tests showed that salicylic acid and benzocaine are percutaneously absorbed from palm stearin to a greater extent than from the other media tested, while the absorption of methyl salicylate from palm stearin is significantly different than from the other three bases.

5.15 Palm methyl esters

Methyl esters can be produced by esterification of the fatty acid with methyl alcohol or by methanolysis of the triglyceride. In certain applications, such as the production of fatty alcohols and alkanolamides, the methyl ester is a preferred starting material to the fatty acid.

The Palm Oil Research Institute of Malaysia is engaged on an extensive project to test the efficacy of palm oil esters as a replacement for diesel fuel. This follows similar tests carried out in other parts of the world.

In Japan the palm methyl ester is saponified to produce toilet soaps of superior colour and odour. In France the same methyl ester is added to laundry soaps as a lime soap dispersant. There are also indications that the sulphonated palm methyl ester is an efficient detergent raw material.

5.16 Textile lubricants

Dihydrogenated tallow dimethyl ammonium chloride $(R_2N^+Me_2Cl^-)$, where R is a long chain alkyl group, is the cationic surfactant that forms the basic ingredient of the home laundry textile softeners that are found on the shelves of the world's supermarkets. The tallow portion of this molecule has effectively been replaced with hydrogenated palm fatty acid, the vegetable origin of palm oil making the product more generally acceptable to some religious communities.

Index

Acid number, 95
Alcohols, 43
Alkaline refining, 47
Anisidine value, 46, 62
AOM stability, 62
Area under cultivation, 4, 5

Bakery products, 78
Biscuit fat, 79
Bleaching, 47, 51, 57
Botany, 12
Broiler chickens, 86
Bunch reception, 29

Cake fat, 79
Carotenoids, 43, 44, 46
Cationic surfactants, 98
Characteristics, 40
Chickens, 86
Chocolate products, 82
Clarification, 34, 36, 37
Cloning, 15
Cloud point, 68
Cocoa butter extenders, 83
Colour, 62, 68, 95
Composition, 39, 40
  see also Fatty acid composition
Confectionary fat 82
Cooking oils and fats 77
Copper 46, 62, 85
Cosmetics 97
Curvularia leaf spot 21

Decanters, 37
Degumming, 47, 48
Density, 40
Deodorization, 47, 57, 59
Detergent fractionation, 65, 67
Digestability, 85
Digestion, 32
Diglycerides, 42, 46
Dimers, 62
Diseases, 21
Distillation, 57, 94
Distilled palm fatty acids, 95
Dough fat, 79
Dry fractionation, 65, 66
Drying, 47, 48

Economics, 22
Elaeidobius kamerunicus, 3, 22
Elaeidobius subvittattus, 22
Elaeis guineensis, 12, 39
Elaeis oleifera, 15, 16, 39
Empty fruit bunch, 87, 89
Environmental factors, 16, 60
Establishment of oil palms, 17
Exploitation, 22
Exports, 1, 8, 9
Extraction, 29, 34

Fats, world production 2, 6
Fatty acid composition, 39, 40, 94, 95
  see also Palm oil, Stearin *and* Olein
Fatty acids, uses, 96
Fertilizers, 19
Filtration, 47
Flash point, 40
Food uses, 71, 74
Fractionation, 62, 95
Free acid, 46, 62, 68, 85
Fusarium oxysporum, 21

Ganoderma, 21
Genetics, 12
Glyceride composition, 41
Glycerol, 94
Gravity settling, 35
Growth of oil palm fruits, 11
Growth rates, 18
Gums, 44

Handling of crude oil, 46
Harvesting, 22, 23, 24
Hydrogenation, 95
Hydroperoxides, 51

Industrial uses, 92
Insoluble impurities, 46
Interesterification, 81
Iodine value, 15, 40, 62, 68, 76, 85, 95
Iron, 46, 62, 85

Lard, 5
Linoleic acid, 15
Lipase, 45
Lipoxygenase, 45

Margarine, 78, 80
Melting, 40
    see also Slip melting point
Metabolizable energy, 85
Methyl esters, 57
Minor components, 42, 64
Moisture, 46, 62, 68
Monoglycerides, 42, 46
Mystrops costaricensis, 22

Neutralization, 47, 48
Neutralization value, 87

Oil content, 23
Oleic acid, 15
Oleic acid distilled, 95
Olein, 64, 68, 73
Organic waste, 20

Palm fruit, 12
Palm mid-fraction, 68, 69, 82
Palm oil
    advantages, 75
    characteristics, 76
    disadvantages, 77
    exports, 1, 8, 9
    fatty acid composition, 76, 85, 93, 94, 95
    imports, 73
    mill effluent, 88, 89
    production, 72, 73
Palm olein, 64, 68, 73
Palm press fibre, 87, 89
Palm stearin, 64, 68, 73, 85
Pastry fat, 79
Peroxide value, 46, 62, 85
Pests, 21
Pharmaceutical applications, 97
Phosphatides, 43, 44, 45
Phospholipids, 43, 44, 45
Phosphorus, 46, 62, 85
Phosphorylase, 45
Physical refining, 46, 47, 55
Pigments, 51
Pigs, 86
Plantation cycle, 24
Planting-organization, 25
Polishing, 47
Pollination, 21
Process control, 57
Productivity, 17
Profitability, 3
Prospects, 7
Purification, 34

Quality, 45, 46

Rapeseed oil, 6, 72
Refined oil specifications, 62
Refining, 45, 47, 52, 53, 61
Refractive index, 40
Ruminants, 86

Saponification number, 95
Saponification value, 40
Shortening, 78
Slip melting point, 62, 68, 76, 85
Soap manufacture, 96
Solid fat content, 42, 68, 76
Soybean oil, 6, 72
Spacing of trees, 17
Specification, refined palm oil, 62
Splitting, 94
Stearic acid distilled, 95
Stearin, 64, 68, 73, 85
Sterilizing, 29
Sterols, 43
Storage of crude oil, 46
Stripping, 32

Table margarine, 80
Tallow, fatty acid composition, 93
Tenera palms, 12, 14, 23
Textile lubricant, 98
Thinning, 19
Threshing, 32
Thrips hawaiiensis, 22
Tin plating, 97
Titre, 95
Tocopherols, 43, 46
Tocotrienols, 43
Totox value, 46, 62
trans Acids, 76
Triglyceride composition, 41
Triterpene alcohols, 43

Unsaponifiable matter, 42, 43
Upkeep, 19

Vitamins, 85

Washing, 47, 48
Weevil, 3
World production, 1, 2, 4, 5, 6, 8, 73

Yield, 3, 4, 5
Yield cycles, 24
Yield forecasting, 24